William Henry Davenport Adams

The Story of our Lighthouses and Lightships

Descriptive and Historical

William Henry Davenport Adams

The Story of our Lighthouses and Lightships
Descriptive and Historical

ISBN/EAN: 9783337251048

Printed in Europe, USA, Canada, Australia, Japan

Cover: Foto ©berggeist007 / pixelio.de

More available books at **www.hansebooks.com**

Ancient Beacon-Fire.

THE STORY OF OUR

LIGHTHOUSES

AND

LIGHTSHIPS

Descriptive and Historical

BY

W. H. DAVENPORT ADAMS

THOMAS NELSON AND SONS

London, Edinburgh, and New York

1891

Contents.

———◆◆———

LIGHTHOUSES.

CHAPTER I.

LIGHTHOUSES OF ANTIQUITY.

POPULARLY, the lighthouse seems to be looked upon as a modern invention,—and if we consider it in its present form, completeness, and efficiency, we shall be justified in limiting its history to the last two centuries; but as soon as men began to go down to the sea in ships, they must also have begun to experience the need of beacons to guide them into secure channels, and warn them from hidden dangers, and the pressure of this need would be stronger in the night even than in the day. So soon as a want is strongly felt, man's invention hastens to supply it; and we may be sure, therefore, that in the very earliest ages of civilization lights of some kind or other were introduced for the benefit of the mariner. It may very well be that these, at first, would be nothing more than fires kindled on wave-washed promontories,

or on hill-tops* commanding clear and extensive views
of the neighbouring waters; but we may be certain
that the difficulties in the way of maintaining such
fires in stormy weather would speedily suggest the
construction of some permanent shelter or protection.
And therefore I see no reason to doubt the statement
that long before Greece had become a maritime nation,
light-towers, which served as landmarks during the
day as well as beacons during the night, had been
planted along the coast-line of Lower Egypt. The
first, however, of which we have any certain information was the celebrated *Pharos of Alexandria,* so
celebrated that its name is very generally applied to
lighthouses in our own day, and exists in the French
Phare. It was erected about 280 B.C., during the reign
of King Ptolemæus Philadelphus, on the strip of shining
calcareous rock lying about a mile off Alexandria that
sheltered both its ancient ports—the Greater Harbour
and the Haven of Happy Return—from the fury of
the north wind and the occasional high tides of the
Mediterranean. Its site was a kind of rocky peninsula
at the east end of the island, and it stood conspicuous
from its great height and the whiteness of the stone of
which it was constructed. Whether the light it supplied was obtained from an ordinary fire, or some more
effective system of illumination, we have no means of

---

* See the passage in Homer ("Iliad," xix. 375):—

" As to seamen o'er the wave is borne
 The watch-fire's light which, high among the hills,
 Some shepherd kindles in his lonely fold."

ANCIENT PHAROS OF ALEXANDRIA.

determining; though we can hardly suppose that to the former Lucan, in his "Pharsalia," would have applied the word "lampada," or that Pliny would have expressed his fear lest, on account of its steadiness ("continuatione ignium"), the mariner should mistake it for a star.

The Alexandrian Pharos was probably destroyed by the Turkish conquerors of Egypt. That it existed in the twelfth century we know from the elaborate description of it recorded by the Arabian writer Edrisi, who says:—"This pharos is without its like in the world for skill of construction and solidity; since, to say nothing of the fact that it is built of excellent stone, the courses are united by molten lead, and so firmly fitted together as to form an indissoluble mass. We ascend to the summit by an internal staircase. About half-way the building becomes much narrower, and from the gallery rises to the top with a continually increasing contraction, until at last it may be compassed within a man's arms. The staircase from the gallery is, therefore, of narrowing dimensions also. In every part it is pierced with windows to give light to persons making use of it." He adds, in language reminding us of Pliny's, that from a distance the Pharos-ray was so like a star which had risen upon the horizon, that mariners, misled by it, steered towards the other coast, and were frequently wrecked upon the sands of Marmorica.

The famous Colossus, or gigantic statue of Apollo,

designed by the sculptor Chares, and erected at Rhodes, has been by some authorities described as a Pharos; but for this statement we confess ourselves unable to find any authority. We pass on, therefore, to the light-tower which the Romans raised on the heights of Dover. In a shattered state, it is still extant; and Mr. Puckle describes it[*] as "a massive shell, the inner face of its walls vertical and squared, the outside with a tendency to a conical form, which was probably at one time much more distinct, allowing for the quantities of external masonry and facing which by degrees must have fallen or been hewn away." The basement alone, however, is of Roman work. The octagonal chamber above was added in the reign of Henry VIII. Its dimensions are about fourteen feet square, and its materials, according to Mr. Puckle, tufa, concrete, and red tile-brick. There is little doubt but that it was raised for the purposes of a Pharos; but as, from its elevated position, it must have constantly been shrouded in mists, its fires were probably discontinued immediately after the Conquest. In course of time it was applied to military uses, the lower chamber being converted into a guard-room; and of late years it has served as a store-house. It is thirty-three feet in diameter, and was formerly about seventy-two feet high.

The first Pharos which performed its duties in a regular manner seems to have been that which Lesches,

* Rev. J. Puckle, "Church and Fortress of Dover Castle," ed. 1864.

the author of the "Little Iliad" (who flourished about the 9th Olympiad), erected on the promontory of Sigeum, at the entrance of the Hellespont. It is figured in the Iliac Table.

Though the most ancient in our records, the honour

THE TOWER AT DOVER.

was not reserved to it of bequeathing its name to its successors, any more than to Columbus the glory of leaving *his* name to the New World. This honour was gained, as I have said, by the mighty tower elevated on the island of Pharos, which served as a model for some of the most celebrated lighthouses erected in later times. Such was the case with the Pharos built by the Emperor Claudian at Ostia, which appears to have been the most remarkable of any on the Latin coast. It was situated upon a breakwater, or artificial island, which occupied the mid space between the two huge moles that formed the harbour; * and its ruins were extant as late as the fifteenth century, when they were visited by Pope Pius II. Not less stately was the Pharos which guided the seamen into the port of Puteoli, the emporium of the foreign trade of imperial Rome; nor that which Augustus erected at the entrance of his new harbour of Ravenna, and Pliny describes with so much enthusiasm; nor that, again, which shed its warning light from the mole of Messina over the whirlpool of Charybdis and the rock of Scylla; nor that which blazed in the island of Capreæ, and was destroyed by an earthquake shortly before the death of Tiberius.

Dionysius of Byzantium† describes a celebrated lighthouse planted at the mouth of the river Chrysorrhoas, where the latter mingles its waters with those of the Thracian Bosphorus (the modern channel of Constantinople). "On the crest of the hill," he says, "the base of which is washed by the Chrysorrhoas, may be seen the Timean tower, of an extraordinary height; and from its summit the spectator beholds a vast ex-

---

* Suetonius, "Claudian," 20.

† Author of an Ἀναπλους Βοσπόρου, *circa* A.D. 190.

panse of sea. It has been built for the safety of the
navigator, fires being kindled for their guidance; which
was all the more necessary because the shores of this
sea are without ports, and no anchor can reach its

A ROMAN PHAROS.
(From a Medal in the D'Estrées' Collection.)

bottom. But the barbarians of the coast lighted other
fires on the loftiest points of the coast, to deceive the
mariner, and profit by his shipwreck. At present,"
adds our author, " the tower is partly ruined, and no
lantern is lighted in it."

Strabo refers in exaggerated terms to a superb light-tower of stone at Capio, or **Apio,** near the harbour of Menestheus—the modern Puerto de Santa Maria. It stood on a rocky headland, nearly surrounded by the sea, and served as a guide for vessels through the shallow channels at the mouth of the Guadalquivir.*

What was the form of the Roman light-towers? This is a question not easily answered, when we remember that Herodian compares them to the catafalques of the emperors. The catafalques were square; but it is certain that quadrangular lighthouses were very seldom constructed. Montfaucon reproduces a medallion, from the famous cabinet of the Maréchal d'Estrées, which represents a Roman lighthouse as a circular tower, built in four stories of decreasing diameter. Another medal, discovered at Apameia, in Bithynia, and also figured by Montfaucon, likewise depicts a circular building. And we shall hereafter see that the Pharos at Dover, as at Boulogne, was of the same form.

One of the most interesting of ancient lighthouses was, unquestionably, the *Tour d'Ordre* at Boulogne,— the ancient *Bononia* or *Gesoriacum,*—erected by the Emperor Caligula for the gratification of a colossal vanity. From the description left by Claude Châtillon, engineer to Henry IV., we learn that it was built about two hundred yards from the brink of the cliff; that its materials were gray and yellow stones and tile-red

---

* Strabo, Edit. Oxon., 1867, p. 184.

bricks; that in shape it was octagonal; that its base measured one hundred and ninety-two feet in circumference; and that each of its twelve stories was a foot and a half narrower than the story immediately below it, so that it assumed externally the outline of a

ROMAN PHAROS.
(After a Medal of Apameia.)

pyramid. In height it was about one hundred and twenty-four feet. As late as the opening years of the seventeenth century three of its stories or vaulted chambers still remained; they were connected by an inner flight of stairs. Even at this period it seems to have been used as a lighthouse; whence, according to a dubious etymology, its ancient name of *Turris ardens* became corrupted into *Tour d'Ordre.*

Along with these ancient lighthouses we may class, I think, the *Tour de Cordouan*, for its origin certainly belongs to a very distant period. The present is the third which has been erected on the same site—a tall cliff dominating the entrance of the Gironde, and washed by the waters of the Gulf of Gascony. The first was built in the thirteenth century, at the solicitation of the merchants of Cordouan, who, as their wealth depended upon the visits of foreign traders, were naturally anxious to remove any difficulties of navigation. A second lighthouse was raised in the middle of the fourteenth century by order of Edward the Black Prince (about 1362–1370). It was forty-eight feet high, and terminated in a platform, where an open wood-fire was kept alive by a holy hermit, who received in acknowledgment of his humane labours a toll of two groats from each passing vessel. It is generally believed that the rock on which the light-house stood was, at that period, a part of the Medoc coast; and this belief is supported by such facts as the similar character of the geological conformation, the depth of the existing channel between rock and main-land, and the nature of the havoc still committed by the sea at Soulac and the Point de Grave.

In immediate neighbourhood to the Black Prince's light-tower was raised a chapel in honour of the

BIRD'S-EYE VIEW OF THE TOUR D'ORDRE OF BOULOGNE.

(From an old drawing by Claude Châtillon.)

The English Coast.

blessed Virgin; and gradually a small village clustered round about, inhabited principally by fisher-folk.

In an old engraving the lighthouse is represented as octagonal in shape, with elongated quadrangular openings; it was strengthened up to its first story with an exterior or second casing of stone.

ANCIENT TOWER OF CORDOUAN.

The present structure was begun—not on the ruins, but by the side, of its predecessor—by Louis de Foix, a Parisian architect, in 1584, and completed in 1600, under his son's direction. It consisted of a circular platform protected by a broad parapet, and of a tower, one hundred and sixty-two feet high, which formed a circular cone, and was divided into four stories, besides the lantern. Each story was of a different order of architecture, and embellished with busts and statues of French monarchs and the Olympian deities. The ground floor was arranged as a spacious four-square vestibule, with four recesses which were used as store-rooms. Staircases in the embrasures of the entrance-gate and the two windows led to cellars and water-supply. The first story, of the same dimensions as the vestibule, but more richly ornamented, was called "The King's Chamber," and opened upon the first outer gallery. The second story was occupied by a chapel, with a fine vaulted ceiling, Corinthian pilasters, decorative carving, and two rows of windows. A bust of Louis de Foix was placed above the chapel door, and a tablet lettered with a quaint inscription in his honour.

This second story was surmounted by a circular pavilion, vaulted, and decorated with composite pilasters, the entablature being crowned by the open balustrade of a second outer gallery which led into the lantern. The lantern was built of hewn stone, and consisted of eight arcades, the piers of which were enriched with columns, while the cupola terminated with a shaft to carry off the smoke of the brazier. In 1727, this stone lantern, which obscured a good deal of the light, was replaced by one of iron ; but the elevation above the sea was not reduced. In course of time,

PRESENT LIGHTHOUSE OF CORDUAN.

however, as more shipping entered the Gironde, this elevation was declared to be inadequate for lighting purposes, and the Chevalier de Borda proposed a scheme for increasing it by one hundred feet. Teulère, the chief engineer to the city of Bordeaux, considered so great an addition unnecessary and dangerous, and submitted plans for an increased elevation of sixty-five feet. These were approved, and in 1788–89 were successfully carried out by their ingenious author; so that the focal plane of the light is now at a height of two hundred and five feet above the ground, and of one hundred and ninety feet above high water.

Teulère's addition is in striking contrast, with its almost barren simplicity and severity of outline, to the richness, grace, and elegance of the Renaissance work of De Foix. Yet the general effect is so impressive as to leave little to be desired; and the spectator gazes with involuntary admiration on the majestic monument of skill and philanthropy which seems to spring so boldly from the heaving bosom of ocean. So Michelet writes in his prose-poem on the Sea:—" During a six months' residence on the coast here, our ordinary object of contemplation—I had almost said, our daily society —was the Tour de Cordouan. We felt very strongly how its position as guardian of the waves, as the faithful warder of the Channel, conferred upon it a living individuality. Erect against the eastern horizon, it assumed a hundred different aspects. Sometimes, in a belt of glory, it triumphed under the sun; sometimes, wan and indefinite, it hovered through the mist, no augury of good. At evening, when it abruptly kindled its red light, and darted forth its fiery glare, it seemed like a zealous custodian, who watched over the waters, impressed and disturbed by his responsibility."

Within the last thirty or forty years a complete restoration of this noble lighthouse has been undertaken, for the purpose of replacing so much of the masonry as had suffered from the weather, and of renewing the sculptures and carvings, which had lost their sharpness. The commonplace buildings erected at different times around the base for the accommodation of the keepers were handsomely reconstructed. And in 1854 it was differentiated from other seamarks by being furnished with a white and red revolving light, visible over a range of twenty-seven miles.

The apparatus of the dioptric system of lighting was introduced into the Tour de Cordouan long ago ; and it was here that Fresnel's earliest experiments were made. For the Cordouan lighthouse has held among the French lighthouses much the same place as the South Foreland among English lighthouses, and has always been chosen as a place for experiments. It was one of the very first in which the *chauffer* was replaced by oil-lamps. In 1782, its lantern was illuminated by at least eighty, each with a metal reflector. And a few years later, when Teulère had suggested to Borda the elements of the catoptric system, the new apparatus was immediately installed here (1790).

INTERIOR OF THE CORDUAN LIGHTHOUSE.

CHAPTER II.

LIGHTHOUSE ADMINISTRATION.

WE may justly claim for our own country the praise, among modern nations, of having been the first to appreciate the full importance of a complete and efficient lighthouse system, and of having made its development and its maintenance on a liberal scale a matter of national concern. But that such should be the case was natural enough, considering the extent of our coast - line as compared with the superficial area of the United Kingdom, the number of our harbours, havens, estuaries, and waterways, the dangers and difficulties of their approaches, and the magnitude of our maritime interests.

The supervision of our lighthouse system is in the hands of three separate boards—one for each of the three kingdoms—and all three responsible to the Board of Trade. These are :—

1. *The Corporation of the Trinity House of Deptford Strond*—for England.

2. *The Commissioners of Northern Lighthouses*— for Scotland ; and

3. *The Corporation for Preserving and Improving the Port of Dublin*—for Ireland.

The history of the Trinity House is but imperfectly known, owing to the destruction of a considerable portion of its archives by fire in 1714. It was founded by Sir Thomas Spert in 1515, and incorporated by Henry VIII. in 1529, under the title of "The Master, Wardens, and Assistants of the Guild, Fraternity, or Brotherhood of the Most Glorious and Undividable Trinity." At first, perhaps, the sole duty of its members was to pray for the souls of sailors drowned at sea, and for the lives of men struggling with the dangers of wind and wave; but at a very early date they were intrusted with the general control of our mercantile marine and the lighting of the English coast. Charters with full powers were granted to them by Elizabeth, James I. (1604), Charles II. (1660), and James II. (1685). It is true enough that signal-lights and beacon-fires already blazed on rocky headlands and at the mouths of our most frequented havens; but as the marine activity of England was rapidly increasing, a more effective and uniform system became indispensable. Moreover, the owners of these private lights and beacons had seldom been animated by patriotism or philanthropy, and levied excessive tolls upon passing ships. To erect and maintain a lighthouse was a profitable speculation; and the privileges conferred upon the Trinity Corporation provoked very general discontent. In the reign of James I. an attack was

made upon them—an attack by no means unwelcome
to the king, for the revocation of these privileges by
the Crown would have placed in its hands the disposal
of a valuable monopoly. The judges charged with the
examination of the Trinity Corporation's claims were
considerably embarrassed by the royal action; and at
length the different parties concerned arrived at a com-
promise, to the effect that the fraternity of the Trinity
House should retain its authority to erect lighthouses,
but that the Crown should enjoy the same privilege
in virtue of the common law. Hence, instead of the
lighting of our coasts remaining exclusively, as Elizabeth had intended, in the hands of the Trinity Corporation, it was divided by the Crown among numerous
private individuals, and the old evil system was to a
great extent re-established, and the privilege again
became such a source of pecuniary advantage that
great exertions were made to obtain it. Even a man
like Lord Grenville could enter in his diary the significant memorandum:—"To watch the moment when
the king is in a good temper, to ask of him a lighthouse." Eventually the system was found to be an
intolerable burden upon navigation. Many of the
lights were shamefully deficient in power; others were
allowed to fall into disuse, and yet heavy tolls continued to be levied. Parliament was constrained to interfere; and in the reign of William IV. an act was
passed which made over all the interests of the Crown
to the Trinity House, and empowered it to buy up the

lights in the hands of private individuals; and as this corporation has usually exhibited no ordinary activity, intelligence, and conscientious zeal for the efficiency of its work, the lighthouse service of England has been brought into a very effective and complete condition.

The story of the two other corporations may be told in a few lines. The Commission of Northern Lights, which is certainly not inferior to the Trinity House in enlightened discharge of its duties, was incorporated by Act of Parliament in 1786, and consists of two magistrates appointed by the Crown, of the sheriffs of the sea-board counties, of the Provosts of Edinburgh and the royal burghs, and the Provost of Greenock.

The lightage of the Irish coast was formerly in the hands of the "Revenue or Barrack Board," but was afterwards transferred, a few years ago, to the Corporation for Preserving and Improving the Port of Dublin, and is now superintended by a body called the Commissioners of Irish Lighthouses. The Barrack Board contented itself with farming the lights out to a contractor, who employed the keepers, and paid all expense of maintenance. "The pay to light-keepers was very small, generally averaging £15 per annum; and as perquisite they had all the unburned portions of the candles, and were allowed to carry on their trades, and keep a public-house in the lighthouse. This was the case at Howth, Wicklow, and Hook Tower (where the keeper was a herb-doctor), and doubtless at other stations." The Irish lighthouse service is now, however, quite adequately organized.

These three boards are all under the general control of the Board of Trade. Before new lighthouses are erected by the Trinity House, they must be sanctioned by the Board of Trade; and prior to the erection of a Scotch or Irish lighthouse, the Trinity House must be consulted, and in the event of a difference of opinion arising, the Board of Trade pronounce their decision, which is final. We think it a regrettable fact that a considerable number of the lights of the United Kingdom should still be under the control of local authorities. The Royal Commission of 1859–60 gave the number as about one hundred and seventy. We believe it has since been reduced; but it is much to be desired that *all* should be brought under the control of one central authority, so as to secure uniformity, efficiency, and economy.

With the structure erected by Winstanley on the Eddystone in 1696 began the great efforts of modern engineering science to direct the powers of nature for the use and convenience of man—efforts continued with brilliant energy and success by Smeaton, the Stevensons, Halpin, James Walker, Sloane, Douglass, and others, who have converted hidden dangers into "sources of safety," and beneficently provided for the mariner's guidance in his trackless path. During the last century an extraordinary increase has occurred in the number of lights, fixed and floating, required to meet the needs of a constantly increasing commerce.

Accurate statistics, however, can be obtained only for the last twenty-five years; but these show that whereas in 1860 the aggregate of coast-lights throughout the world did not exceed eighteen hundred, the present total does not fall short of about four thousand two hundred.

COMPARATIVE STATEMENT

Of the Coast-lights in the chief countries of the world (exclusive of their outlying Possessions) in the years 1860 and 1885 respectively.

| | LIGHTHOUSES. | | | | | | | | | LIGHT-VESSELS. | | |
| | 1860. Number. | | | 1885. Number. | | | Increase. Number. | | | 1860. Number. | 1885. Number. | 1885. Increase. |
COUNTRY.	1st Class.	Secondary.	Total.	1st Class.	Secondary.	Total.	1st Class.	Secondary.	Total.			
England and Wales	24	178	202	43	296	339	19	118	137	42	57	15
Scotland	17	112	129	23	166	189	6	54	60	1	4	3
Ireland	11	74	85	19	108	127	8	34	42	5	11	6
United Kingdom	52	364	416	85	570	655	33	206	239	48	72	24
United States	26	314	340	51	1917	1968	25	1603	1628	39	23	*16
France	32	193	225	39	374	413	7	181	188	3	9	6
British America	4	87	91	5	631	636	1	544	545	1	15	14
Sweden and Norway	3	115	118	8	321	329	5	206	211	2	8	6
Italy	3	88	91	16	234	250	13	146	159	..	13	13
Russia	2	63	65	14	164	178	12	101	113	12	16	4
Australia	6	33	39	24	231	255	18	198	216	5	14	9
Austria	..	10	10	2	61	63	2	51	53	..	..	..
Denmark	2	68	70	7	45	52	5	123	†18	7	11	4
Spain	9	41	50	4	167	178	2	126	128	..	..	..
Netherlands	3	55	58	8	94	102	5	39	44	..	3	3
India	..	42	42	13	74	87	13	32	45	7	9	2
Germany	1	31	32	10	193	203	9	152	161	8	22	14
New Zealand	..	3	3	6	66	72	6	63	69	..	2	2
China	..	4	4	14	41	55	14	37	51	1	13	12
Turkey	1	13	14	..	..	129	..	..	115	1	5	4
Japan	..	..	..	8	49	57	8	49	57	..	2	2
Brazil	..	16	16	9	47	56	9	31	40	..	1	1
Portugal	1	14	15	1	29	30	..	15	15	..	..	..
Belgium	1	5	6	1	21	22	..	16	16	2	3	1
Greece	..	..	..	3	54	57	3	54	57	..	1	1
Totals	146	1559	1705	335	5383	5847	190	3827	4132	136	242	122

* Decrease, owing to the large substitution of permanent lighthouses for lightships.

† Decrease, owing to the cession of Schleswig-Holstein to Prussia in 1864.

The preceding table, prepared by Sir James Douglass, and submitted to the British Association in 1886, illustrates the relative progress of each of the chief maritime countries in the extension of their system of lighthouses and light - vessels between 1860 and 1885. It will be seen that Japan, which had not a single coast-light in 1860, has now fifty-seven (of which eight are first-class lights) ; while China, which had only four secondary coast-lights, has now fifty-five (fourteen of these being first-class). The greatest increase, however, is found in British America, where in 1860 were only ninety-one coast-lights, but in 1885 no fewer than six hundred and thirty-six.

We may note, in conclusion, that the coast-line of England measures 2,405 nautical miles, that of Scotland 4,467 that of Ireland 2,518, and that of France 2,763 nautical miles. In 1885 England had 339 lighthouses and 57 lightships; Ireland, 127 and 11; Scotland, 189 and 4,—against France, 413 and 9. The proportion of lights to the coast-line was : for England, 1 to every $6\frac{29}{395}$ miles; for Ireland, 1 to every $18\frac{12}{66}$ miles; for Scotland, 1 to every $23\frac{28}{195}$ miles; and for France, 1 to every $6\frac{33}{66}$ miles. The advantage, therefore, still rested with England, though France had made an immense advance in the last quarter of a century.

It may be added that the French lighthouse service is administered by the Board of Bridges and Highways (*Conseil Général des Ponts et Chaussées*), which is composed of officers of the marine, hydrographic engineers, members of the French Institute, and other persons acquainted theoretically or practically with the sciences bearing upon navigation. The executive is committed to the Inspector-General "des Ponts et Chaussées," who has under his orders a certain number of engineers in each maritime district, charged with the supervision, construction, and maintenance of the lighthouses. This bureau has its special manufactories in Paris, where experiments are tried with illuminating apparatus, and the artisan is trained in the scientific principles on which they are based, in the calculation of angles, curves, prisms, lenses, and the like. This centralization insures not only efficiency but economy, the whole cost of the French system not exceeding £40,000 per annum. Finally, to France, as to the United States, belongs the honour of having regarded the lightage of her coasts, not as a source of public or private profit, but as a work of humanity. It is to be hoped that Great Britain will follow so excellent an example, and before long abolish the tolls now levied upon shipping for the maintenance of her lighthouses; for though these have been subjected to a large reduction, they remain as a burden upon commerce, and a burden which commerce ought not to bear.

CHAPTER III.

GEOGRAPHICAL DISTRIBUTION OF LIGHTHOUSES.

WHEN the military protection of our island
shores is discussed, the civilian is strongly
warned of the necessity of maintaining more than one
" line of defence." We may confidently assert that a
similar necessity exists in connection with their complete and satisfactory lightage. We know, too, that in
the construction of these lines of defence a great variety
is observed : that at one point a battery is erected, at
another an ordinary earthwork, at a third the most
complicated system of bastions, ravelins, and redoubts
which military engineering can devise. A similar
diversity is introduced into the disposition of those
defences which are erected in the interests of commerce. Tracing the long coast-line of our country, we
perceive considerable differences not only in the situation of our lighthouses, but in their mode of construction, their height, their illuminating apparatus. Some
are planted on lonely rocks far out in the waste of
waters ; others on wind-swept headlands or low sandy

spits, where at night the sea-birds dash against the lighted pane; others at the entrance to noble havens in which the tallest ships may ride in peace. One is a shapely tower, like a pine tree, springing from the very bosom of the waves; another, a square, castellated structure like the keep of a feudal castle. One exhibits a fixed light with the steadfast glow of a star of the first magnitude; another suddenly leaps out of the womb of darkness, and throws over the sea its arrow of flame, to fade away and reappear a few moments later with the same strange and impressive lustre. One is visible at a distance of twenty-seven miles; the range of illumination of another is only five. Nor are all lights of the same colour. Some are red, with an intense ruby-like splendour; others, white; others, blue or green. This variation in aspect, range, and position has, like the variation in aspect, range, and position of our forts and batteries, a special object.

The system of lightage in general adoption surrounds the coast with three lines of defence. The outermost of these is formed of lighthouses with a very extensive range—lighthouses of the first class, which are planted upon reefs and islets some miles out at sea; or on the summit of capes and promontories, exposed to the full fury of the gale; and along our British shores these are so amply provided that it is impossible to approach any important part without sighting one or more of them, such as the Longships, the Wolf, the Eddystone, the Bell Rock, the South Stack, the Skerryvore. Inside

these, the navigator comes upon a second circle, composed of lighthouses of the second and third class,
which indicate the position of shallows and sandbanks,
and, more particularly, the safest channels in the mouth
of a river or the entrance of a port. And, lastly, when
the ship reaches the wished-for haven, she is guided to
her moorings by the " harbour lights " which cluster
about quay, pier, or breakwater. The lighthouse map
attached to this volume will show that, so ample is the
illumination of our shores, the luminous circles created
by our " fire-towers " are not simply contiguous, but
actually overlap each other, maintaining an unbroken
belt of light.

It has even been contended that the British coast is
too plentifully supplied with those beacon-marks, and
that in their very abundance lurks danger. Such might
very probably be the case if the means had not been
provided of easily distinguishing between them—means
so simple and yet so effective that it is almost impossible for the mariner to experience any difficulty in
differentiating them. By day the distinction presents
no difficulties ; only at night could mistakes be made ;
and to obviate them—and, if by accident the navigator
has been driven out of his reckoning, to enable him at
once to calculate his true position—a distinction is made
in the character of each light, which, as we have said,
is a safeguard against danger. Broadly speaking, all
lights come under two classes : they are either fixed—
that is, they undergo no alteration ; or they are not

fixed. The former can be differentiated by the use of coloured glass—white, red, or **green; or** two fixed lights **may be placed** together, **or** superposed—two white lights, or **two** red, or red and white, or red **and green :** the second class can be **varied** more effectively. **For** instance, there can be the " **revolving** light," **in which** the rays are **separated** by dark **intervals, and as they** travel round, illuminate the horizon **in due** succession. Now this kind of light is also susceptible **of variation. The dark intervals can be lengthened** or shortened : **the** ray can **come** round in half a **minute or a minute or** a minute **and a** half; **or** coloured shades **may be introduced, and made** to **revolve** alternately **with the** white rays; **or two** coloured rays may be shown to one white ; and so **on.** Other changes **will be** readily ima-**gined by the** reader.

The late Mr. Robert Stevenson proposed three more **variations, all** of which are now in use—the " flashing " light, the " intermittent," and the " double lights in one tower." The " flashing " light is **one** which shows two flashes and two eclipses, or more, in a given period. These flashes differ from the paling and increasing rays of a revolving light in **being more** sudden in their **action and briefer** in their duration. **They cannot be** modified very easily by time-periods, **as it is not ad-**visable **to** separate the flashes by **any** long interval of darkness, but **coloured** glasses can again be introduced, and a white and red flash, or red and white, be brought into immediate contrast. Recently an important differentiation has been brought forward in Mr. Wigham's "group-flashing" system—two, three, or four, or even five flashes appearing in such swift succession as to form a distinct group, which is separated from the next group by a dark period sufficiently long to be distinguished from the momentary intervals between the flashes. A very large number of combinations and permutations may be effected in this system. The flashes may be lengthened or shortened; so may the dark intervals; coloured flashes may be used, or coloured and white flashes alternately; and so on.

We now come to "occulting" or "intermittent" lights, in which the leading feature is the *disappearance* of the light, through the automatic and regular interposition and withdrawal of a screen in front of it, or the periodic lowering or raising of a metal cylinder round the chimney of the lamp. Here, again, a variation is possible in the duration of the intervals of light or darkness. The light period may be thirty seconds, and the dark thirty seconds; or twenty and forty seconds respectively. Or there may be a sudden eclipse every half-minute for three seconds. There may also be a change of colour, the red light being occulted at one time and the white at another. But this class of lights is not much used, and, we believe, is not in favour with our seamen, who have been habituated by long experience to watch for the ray or flash, and *not* for the interval of darkness. In distributing these different kinds along the coast, the rule observed by our engineers is, if possible, **never to place two similar** lights within a hundred miles of each **other.** Taking a **portion of** the south coast of England, **from the** Land's **End** to Start **Point,** let us note **the varia-tions** introduced :—1. White **with** red sectors, intermittent ; 2. White and red alternately, revolving ; **3.** White **and** coloured sectors, fixed ; **4. White,** fixed ; 5. Double **light** in same tower, white, revol**ving** twenty **seconds and fixed ; 6.** Green, **fixed ; 7. Red, fixed ; 8.** Double **light in same tower, white, flashing and fixed ; 9.** Double **light in same tower, white, occulting and fixed ; 10. White,** occulting ; **and 11. Double light in same tower,** white, revolving, one minute, and fixed. **It will** be **seen that each of these** eleven **lights** is distinctly differentiated **from the others.**

An **important service is performed by coloured lights,** to which reference **must briefly be** made. **The main light,** whatever **may be its individual** character, **is de-signed to do what may be** called " distant **work. It lifts its head high," says** Mr. Edwards, **" reaches out across the waters as far as the** horizon, **and, we may fancy, strives to look over** and beyond the **earth's** curve, **so as to** catch **the eye of any** far-off struggling mariner **who may** need **its** guiding **light." But " it** is the practice **at many** places **to throw** from **the light-house a** subsidiary **light** of a special character, intended more **particularly to mark any rocks or shoals** in **the** immediate **neighbourhood. On** seeing this **special** light **sailors know that they are in** danger, and by its bearing are assisted in shaping a course of safety. Neither long ranges nor powerful lights are required in these cases......The portions of the main light appropriated are generally given a special character by means of colour, and are called *sectors*, the object being practically achieved by simply causing the light going out in the direction of the dangers requiring to be indicated to pass through a vertical strip of coloured glass. Red is the colour mostly employed; but it is plain that any kind of sector differing from the character of the main light will be applicable."

At present not fewer than eighty-six distinctive characters are in use throughout the lighthouses and light-vessels of the world; and this important question of securing for each light within a certain definite range a complete distinctive individuality is receiving the careful attention of lighthouse authorities both at home and abroad.

CHAPTER IV.

A N illuminating apparatus, of whatever material or on whatever principle it may be constructed, is placed in the lantern of a lighthouse in order that it may so inflex the rays, which would otherwise (and naturally) proceed in straight lines, that instead of being thrown upon the sky and thereby expended uselessly, they may be made to fall upon and light up certain points at sea. And if the light is to be fixed, and intended to be visible all round, and from the horizon to the base of the tower, the upper rays issuing from the illuminating apparatus must be directed downwards and the lower upwards, so as to increase the illumination. If it be simply desired to illuminate a narrow strip of sea, extending from the horizon to the base of the lighthouse, all the rays must be inclined laterally, or they may be concentrated and brought to bear upon particular objects.

Originally two systems were employed in order to accomplish these results. The first, called the *catoptric*, depended on the use of silvered parabolic reflectors; the second, called the *dioptric* (or *refracting*) employed lenses of a peculiar construction. After a while these two systems were combined, first in the ordinary *cata-dioptric,* and second in Mr. Stevenson's ingenious *holophotal* arrangement, which is applicable either to the one or the other.

Before entering upon an explanation and description of these systems, I shall attempt to furnish the reader with a brief historical sketch of the progress of light-house illumination.

The earliest sources of light made use of were wood and coal fires, which blazed or flickered, according to the condition of the atmosphere, on the top of massive beacon-towers erected at conspicuous points along the coast, or at the mouths of ports and harbours. Such fires must have been very uncertain and even mis-leading guides; yet so slow was the development of better modes of illumination, that as late as 1822 a coal fire was maintained at St. Bees lighthouse, on the Cumberland coast, and was only then replaced by catop-tric oil-light apparatus. For nearly two centuries these bonfires prevailed; some of them being closed in with bars, and so made to keep their glow towards the sea, while the landward side was open, and others being en-closed in glazed lanterns. Candles were introduced as illuminants towards the close of the seventeenth cen-tury, as in Rudyerd's Eddystone lighthouse, and con-tinued to be in vogue until far on into the eighteenth.

In 1759 Smeaton's Eddystone was lighted by twenty-four tallow candles, weighing two-fifths of a pound each. From experiments made by Sir James Douglass, it appears that the intensity of the light of each candle must have been about 2·8 candle units each, so that the aggregate intensity of radiant light from the twenty-four candles did not exceed sixty-seven candle units. The first attempt towards improvement was the employment of a lantern of eight sides, the centre of each being occupied by a huge semi-convex lens of greenish glass, of about twenty inches in diameter and nine inches in thickness. It was supposed that this would intensify the rays proceeding from candles placed in the focus; but as the candles burned rapidly, it must have been doubtful whether they were at any time in focus at all. A small disk of tin was next adopted; it was shaped nearly to a parabolic curve, and lined with plaster, which again was lined with small pieces of ordinary looking-glass, and in this way an attempt was made to form a parabolic reflector. As late as 1852 the illuminating apparatus at Portpatrick lighthouse was of this primitive kind. "It consisted of mirror glass fixed in parabolic moulds of plaster as reflectors, six lamps with broad wicks, but without glass chimneys; and the glazing of the lantern, consisting of small panes of common window glass, with a mass of woodwork framing, forming a very inferior light."

About the end of the eighteenth century oil-lamps

were brought into use. These were provided with broad wicks, and furnished a small proportion of light in comparison with their immense volume of smoke. Already, in 1763, a master mariner of Liverpool, named William Hutchinson, had introduced there a catoptric apparatus; but the first great step in advance was taken in 1782, when Argand invented the cylindrical wick-lamp. About the same time Teulère, a French engineer, accomplished a signal improvement in the reflectors, which he made of mirrors of perfect polish and a better shape. And by causing these mirrors to revolve around the focus of illumination, so as to project successively towards different points of the horizon the collected ray of light, he invented "the eclipse." Teulère's invention was first applied at Dieppe,* where the celebrated Borda caused a small revolving apparatus of five parabolic lenses.

In 1825 the French lighthouse authorities accomplished another very important improvement by introducing the dioptric system of Fresnel in conjunction with Arago and Fresnel's improvements on the Argand lamp by the addition of a second, third, and fourth concentric wick. It was adopted in England on July 1, 1836.

In 1827 coal gas was employed as an illuminant at the Troon lighthouse, between Ardrossan and Ayr; and in 1847 at the Hartlepool lighthouse, Durham—the latter

---

* There is some evidence, however, that a small revolving apparatus, with three reflectors, was in use at Marstrand, in Sweden, before this date.

combining it with a Fresnel first-order dioptric apparatus. The costliness of gas, where it had to be manufactured in small quantities and at isolated stations, prevented its general adoption. In 1839 experiments were made at the Orford Low Lighthouse, Suffolk, with the Bude light* of the late Sir Goldsworthy Gurney, consisting of two or more concentric Argand gas-burners, superimposed, which produced a brilliant rose-coloured light, through the action of oxygen gas on a flame derived from the combustion of fatty oils. Its expense, however, proved a fatal drawback. Electricity was next invoked as an illuminant, and in 1857, at the suggestion of the illustrious Faraday, the Trinity House experimented at Blackwall with one of Professor Holmes's direct current magneto-electric machines, and with so much success that in the following year, at the South

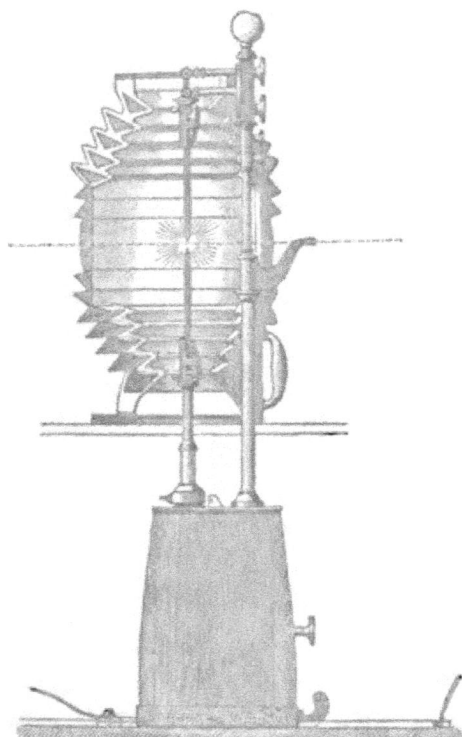

ELECTRIC APPARATUS FOR A
FIXED LIGHT.

* So called from Bude, in Cornwall, the inventor's residence.

Foreland High Lighthouse, two machines were installed, and the electric arc light formally established as a rival to oil and gas for lighthouse illumination. In 1859 the experimental trials at the South Foreland were discontinued; but in 1862 the electric light was permanently applied at the Dongeness or Dungeness lighthouse. In the following year it was introduced by the French authorities at Cape la Hève, where it furnishes a ray equal in intensity to five thousand Carcel burners, and visible for upwards of twenty-seven miles.

In 1871 Professor Holmes having perfected a new alternating current machine, two were applied to the new lighthouse on Souter Point, Durham, and in 1872 to both the High and Low Lighthouses at the South Foreland, where they are still in successful operation. According to Sir James Douglass, the early experience with the electric light at Dungeness was far from encouraging. "Frequent extinctions of the light occurred from various causes connected with the machinery and apparatus, and the oil light had, at such times, to be substituted. As no advantage can counterbalance the want of certainty in signals for the guidance of the mariner, no further step was taken in the development of the electric light until the latter part of 1866, when favourable reports were received from the French light-house authorities of the working of the Alliance Company's system at the two lighthouses of Cape la Hève. Complaints were also received from mariners, in the locality of Dungeness, of the dazzling effect on the eyes when navigating, as they are there frequently required to do, close inshore, thus being prevented from rightly judging their distance from this low and dangerous point. Therefore in 1874 the electric light was removed from Dungeness, and a powerful oil light substituted." In 1877 the electric arc light was installed at the Lizard lighthouses on the south coast of Cornwall, and it has since been introduced at St. Catherine's, Isle of Wight, and at the High Tower on the Isle of May, Firth of Forth. The machines now in use at Souter Point and the South Foreland are the "alternating current" machines.

Experiments were made with various dynamo-electric machines at the South Foreland in 1876, in order to ascertain which was best fitted for adoption at the Lizard. The one selected was the direct-current machine invented by the late Dr. Siemens, and two machines of this type were set up at the Lizard in 1878. But as they failed to work with due regularity, and as satisfactory reports were received of the success of a very powerful alternating-current machine invented by Baron de Maitens of Paris, one was ordered and sent to the Lizard for trial, where it has ever since worked in the most satisfactory manner. The experience thus acquired led to the adoption of the De Maitens machine at St. Catherine's Point; but it was first despatched to the South Foreland, where experiments were then being made as to the relative illuminating value for lighthouse purposes of gas, oil, and electricity.

In 1862 the lime-light invented by and named after Lieutenant Thomas Drummond—which is produced by the combustion of oxygen and hydrogen on a surface of lime—was tested at the South Foreland, but not with encouraging results.

The rapid development of the electric arc light as a lighthouse illuminant naturally provoked the emulation of the partisans of gas, and in 1865 the attention of all concerned was directed to the gas system of Mr. John R. Wigham of Dublin—a system which is now the only direct and successful rival of electricity. Mr. Wigham having been requested by the Commissioners of Irish Lights to report on the subject of oil gas for lighthouses, found a difficulty in getting hold of a suitable burner, and was led to invent for this purpose an ingenious combination of fish-tail jets in what is called the Crocus Burner, which, by the introduction of an oxidizer of mica or other material, renders the flame not only smokeless, but exceedingly white. The combustion is also assisted by a bottom cone for equalizing the current of air to the flame. And further, this burner requires no chimney-glasses, the breaking and cleaning of which is often the cause of much inconvenience in lighthouse maintenance. When photometrically tested, the illuminating power of this burner was proved to be two and a half times that of the largest and most powerful fan-wick oil-lamp hitherto used in the lighthouses of England or Ireland, while its cost for gas consumed was much less than that of the oil-lamp. The Commissioners were so satisfied with this result that they ordered the construction of a gas apparatus at the Howth Baily lighthouse, where the new illuminant was first exhibited in October 1865, and has continued without intermission up to the present time. The gas was for a short time made from oil, then from shale, and latterly from rich cannel, as producing gas of sufficiently good quality.

On the general subject of gas in lighthouses we quote Mr. Wigham's remarks :—" A flame obtained from gas by the method above described possesses in an eminent degree both quantity and intensity. It is unnecessary to say that the former quality is of great importance in lighthouse illumination. Every one knows that the lime light and the electric light, while possessing exceeding intensity, are deficient in quantity, and we are all familiar with the deep darkness to be found outside the line of the rays thrown out by these lights. This want of divergence is a serious drawback to their usefulness, as it necessitates exceeding accuracy of adjustment in the necessary lenticular apparatus......One manifest advantage possessed by the gas over the electric light is the greater simplicity and much smaller cost of the apparatus required for its generation, and the great ease and certainty with which gas can be made and its light maintained. The simplicity of the gas apparatus is such that any ordinary labourer can manage it; but in the case of the electric light skilled labour is required to superintend the electro-magnetic machines from which the electricity is produced, and the steam-engines by which these machines are driven. To guard against any breakdown in the electric light, duplicate steam-boilers, duplicate engines, duplicate electrical machinery, and duplicate electric lamps are provided, and all this is not only very complicated, but enormously more costly than the most perfect gas apparatus."

A more powerful light being still desirable, Mr. Wigham designed another form of burner, consisting of rings of double gas-jets, which can be lighted or cut off according to the requirements of the weather. Professor Tyndall, who examined it at the Howth Baily lighthouse in June 1869, describes it as consisting of a series of concentric fish-tail jets. The three central rings embrace a group of twenty-eight jets, and this is the light employed under ordinary circumstances at Howth Baily. To this central group can be added in succession four other circles of burners, each embracing twenty jets. Thus the lowest light employed is emitted by twenty-eight jets, the next in power by forty-eight, the next by sixty-eight, the next by eighty-eight, and the next by one hundred and eight jets of gas. It is possible, therefore, to employ lights of five different powers.

A notable advantage of the gas system is that the gas-burner while lighted is almost independent of all care on the part of the light-keeper, and cannot, except intentionally, be burned so as to give *less* than its proper focal size of flame. The oil-lamp, on the contrary, requires continual watchfulness; and although a careful light-keeper will endeavour to keep its flame at full focal height, yet there is no doubt that it requires constant skill and attention to do so, and that sometimes a proper light is *not* maintained. Another peculiarity of the gas light is, that when it is desirable to extinguish and relight its flame for any purpose, as in the case of intermittent lights, or when additional power for penetrating fogs is required, the changes can be effected in a moment. This cannot be done satisfactorily with oil or paraffin lamps, for any manipulation of the wicks causes incrustations of carbon upon them, irregularity of flame, smoke, and consequent loss of light. The gas system has also the great collateral advantage that a supply of gas at a lighthouse is a motive power always at hand, by which at a few minutes' notice fog trumpets or whistles may be sounded. This has been carried out at the Howth Baily, where gas has been employed to drive a Hugon's patent gas-engine, which requires only the application of a match to set it in motion. Many interesting experiments have been made there with gas as a means for fog-signalling. An engine for compressing air, and thus sounding an enormous trumpet, was driven by gas, and quite recently *guns* charged with and fired by gas have been tried with good effect.

Great economy necessarily results from the use of gas for "intermittent" lights; for if the light be intermitted with equal periods of darkness—for example, three seconds of light followed by three seconds of darkness—it is evident that about one-half the quantity of gas which would otherwise be expended will be saved. It became therefore a desideratum to apply it to revolving lights; and though this was no easy task, Mr. Wigham after a while accomplished it by making use of the catoptric or reflector system. A mechanical arrangement enabled him to apply gas to a series of revolving reflectors: "a ground-in" gas-joint, fed from a central source, and furnished with radial arms to carry the gas to the reflectors, admits of its being burned in the focus of each reflector with perfect steadiness, while the revolution of the reflectors brings in turn the face of each towards every part of the horizon.

The annular lens, of which eight are generally used for revolving lights, not only transmits the whole of the rays horizontally, but also collects them into one parallel beam. It will be evident, therefore, that the power of the annular lens is greatly superior to that of the refracting belt (used for fixed and intermittent lights): it has been ascertained by calculation to be thirteen times more powerful. That this enormously greater power can be availed of only in revolving lights is obvious; for the fact of the horizontal rays being parallelized of necessity causes angles of darkness, and

to give light to those dark **portions of the** horizon it is
necessary **to cause** the octagon **of** lenses to **revolve so**
that the beam from each lens may traverse **the whole**
horizon.   **Eight** portions of the **horizon are illuminated**
by the beams, but the **eight** portions **between them**
(which **are of much** greater extent) **are in** darkness.
It will be readily seen how advantage can be taken of
the economy **of** intermittent gas-flashes for revolving
lights without depriving the mariner of any light. We
simply **make use of** the periods **of** darkness.   Thus,
suppose the apparatus **set in motion,** what we may
call beam No. 1, now lighting **a** certain point of **the**
horizon, will move on **till it occupies the** place now
held **by beam No. 2;** beam **No. 2** will replace beam
No. 3; beam No. 3 will replace No. 4; **No.** 4 will
replace **No. 5,** and so on, **beam No. 8** replacing beam
No. 1.   Thus every part **of the horizon** will have been
visited by a beam or flash of light, and if the complete
revolution of the octagon of lenses occupies say ninety-
six seconds, a beam **or** flash of light will reach **an**
observer stationed **at any point** of the horizon every
twelve seconds; but if, **when beam No.** 1 reaches the
place occupied by beam **N**o. 2, the revolving machinery
is stopped, and remains stationary for twelve seconds,
during which time the gas **is** extinguished and not
re-lighted **till** the machine is again set in motion, it is
evident that **the observer** will receive a flash every
twenty-four seconds instead **of every** twelve.   As,
however, it is not convenient to stop **the** revolution of

the octagon of lenses—which is very heavy, and could not soon regain its proper velocity—we can accomplish the same end without stopping the machinery by extinguishing the gas at the moment when the whole horizon has been illuminated, and not re-lighting it till the lapse of the same period of time as was occupied by that illumination. Thus, say an observer has received a flash from the lens of beam No. 4, beam No. 3 is coming on, and under ordinary circumstances he would receive a flash from it; but the moment the light reaches him (perhaps showing him a momentary glimmer) the gas is lowered, and not raised again till lens No. 2 is opposite to him, when he receives the flash exactly twelve seconds later than he would otherwise have received it. Thus, in the case we have been supposing, every alternate lens passes over one-eighth of the whole horizon in the dark, and the same effect is produced as if the machine were stopped. It would seem, therefore, that this plan of simply doubling the duration of each dark interval must necessarily reduce the consumption of gas by one-half.

Another invention by Mr. Wigham (whose own descriptions of his apparatus we are freely using) which has come into extensive use is that of "the group-flashing light," or, in other words, a continual intermittent light in conjunction with revolving annular lenses. Instead of extinguishing the gaslight during the passage of every alternate lens, as just described, a little bit of wheelwork connected with the revolving machinery is so arranged that the gaslight is continuously lowered and raised again without interrupting the ordinary revolution of the lens. The special advantage of this device, that the constant extinction and re-exhibition of the light powerfully arrests the attention of the mariner, must be insisted upon. The name "group" is applied to it because, instead of a single flash, a cluster of six or eight flashes reaches the eye at each recurring period. The required duration of the individual flashes composing each group is secured by regulating the revolution of the lenticular apparatus to the necessary speed. Professor Tyndall has reported upon this invention in very favourable terms. He describes it as producing a powerful and indeed splendid effect, and as giving a revolving light so distinctive a character as to render it perfectly unmistakable.

A very important improvement in lighthouse illumination has been effected by Mr. Wigham's triform and quadriform system of gas-burners. Formerly only one central illuminant was used in first-order lighthouses, the dioptric apparatus for which consisted of three parts—the great central annular lens, the top prisms, and the bottom prisms. The light from these top and bottom prisms being very feeble—not more than one-fifth of the whole—Mr. Wigham considered that it would be better to use three of these central lenses superposed, with a light in the focus of each, thus securing an addition of light in the ratio of 240 to 100. The three superposed lenses do not take up more space

than the original dioptric apparatus, so that the same
lantern will contain them; and the extra cost of consumption is very small, for only one out of the three
burners is used in clear weather, the others being added
when the weather is thick and foggy.

The question, which is the best lighthouse illuminant—
oil, gas, or electricity? has been under the consideration
of a strongly constituted committee, appointed by the
Board of Trade; but its decision was not so clear as could
have been desired. So far as we can judge from the
evidence before us, gas seems specially adapted for use
in the more important shore lighthouses, and has unquestionably been very successful at Gally Head and
Howth Baily. For rock lighthouses it is probable that
oil will continue to be employed. The electric arc light
finds many powerful advocates, and has been installed,
as we shall see, at several first-class English and Scotch
stations; but there seems a consensus of opinion among
seamen that in foggy weather it is ineffective, and in
absolute intensity and extent of range it is at present
surpassed by the triform and quadriform gas-burner
system. The subject is one of much interest and high
importance, and we wish the committee had been able
to give a more conclusive deliverance upon it. Meanwhile, let us bear in mind the fact that the practical
application of electricity has by no means attained to
its full development, and that its future cannot possibly
be forecasted.

The oil formerly used in lighthouses was spermaceti,

but since **1846** colza has **been generally** adopted. The substitution **of** the **new** mineral oils for colza was considered **as far back** as 1861; **but** they **were then** so dear, and so imperfectly refined, **that the** lighthouse **authorities** decided against **them.** In 1869, **however,** the price **of mineral oil, of good** illuminating quality and safe flashing point, having **been reduced to about one-half of** that of colza, the **Trinity House,** after **a series of** satisfactory experiments, recommended the change. **It was** found **during these** experiments that "the improved combustion effected in the colza burners, **in their** adaptation for consuming **mineral oils, had the effect** of increasing their mean efficiency, **when** burning colza, **45¾ per** cent. A further **advance was** made **during** these experiments **by** increasing the **number of** wicks **of the** first-order burner **from four to** six, more than doubling **the intensity of the** light, while effecting **an** improved compactness **of the luminary per unit of** focal area of 70 per cent." **The value of** a light given **by a six-wick** burner equals, it is said, that of seven hundred and twenty-two sperm candles, while the **old** four-wick burner **gives a light equal only** to that of three **hundred and twenty-eight** sperm candles. Sir **James** Douglass, **a few years** ago, introduced a seven-wick burner, with proportionately greater results. It is **available of course** either **for colza or** paraffin. When **burning** the former, **the** wicks are raised about a quarter **of an inch** above **the top of the** burner, **and the oil allowed** a slight **overflow;** but in burning **paraffin, the** wicks are not raised more than one-sixteenth part of an inch, and the oil is kept down about an inch and a quarter below the burner's edge. The consumption of oil, whether colza or paraffin, for a four-wick burner is about eleven hundred gallons, for a six-wick burner seventeen hundred and fifty gallons, yearly.

We come now to an explanation of the *catoptric* and *dioptric* systems of illuminating apparatus. Until within the last few years the catoptric system was almost exclusively employed in our English lighthouses, but it has now given way in an extensive degree to the dioptric.

In the accompanying illustration we represent a plan and elevation of

CATOPTRIC APPARATUS.

a catoptric apparatus, consisting of nine Argand lamps and cup-shaped metallic reflectors, arranged in groups of threes, and set in motion by an iron framework mounted on a spindle. Eclipses at longer or shorter intervals are produced according to the rate of speed at which this spindle rotates. If desired, the apparatus may have *four* vertical faces, and the lamps affixed to each may be three, five, seven, or ten. At Beachy Head there are three faces, each face fitted with ten lamps and reflectors, so that thirty lamps and reflectors combine to furnish a ray of light of great intensity.

The form of the reflector is that of a parabola, which has been found the best for reflecting forward in a parallel beam all the rays from the lamp which strike its surface; and the lamp or burner is carefully adjusted so as to insure that the flame shall be exactly in the focus of the reflector, and thus all waste of light prevented.

The obvious objections to this system are, that it necessitates a vast amount of trouble in cleaning and polishing, and that the number of lamps in use entails the consumption of a large quantity of oil, and develops a considerable and inconvenient amount of heat.

The reflectors used for this system are made of sheet copper, plated in the proportion of six ounces of silver to sixteen ounces of copper.* They are shaped into a paraboloidal form by a laborious and difficult process of beating with various kinds of mallets and hammers; and when thus shaped are strengthened round the edge by means of a strong bezel, and a strap of brass attached to it. Polishing powders are then applied, and the instrument receives its latest finish.

To test the shape of the reflector, two brass gauges are employed. One is for the back, and used by the

* Stevenson, "On Lighthouses," pp. 92, 93.

workmen during the process of hammering; the other, while the mirror undergoes the final touches, is applied to the concave face. Its reflecting power is then proved by trying a burner in the focus, and measuring the intensity of the light at various points of the reflected conical beam.

The flame used in a catoptric apparatus is generally derived from an Argand lamp, with flat wicks an inch in diameter. Sometimes the burners are tipped with silver to prevent the wick from being wasted by the great heat evolved. They are also fitted, in some Scotch lighthouses, with an ingenious sliding apparatus, by means of which they can be removed from the interior of the lantern at cleaning time, and afterwards replaced with facility and exactness.

Catoptric lights may be arranged into nine distinct classes: fixed, revolving white, revolving red and white,

AN ARGAND FOUNTAIN LAMP.

revolving red with two whites, revolving white with two reds, flashing, intermittent, double fixed, and double revolving white.

In addition to what we have already said of these and other distinctions, we may here quote Mr. Stevenson's definitions:—

The *fixed*, he says, maintains a regular and steady appearance. The reflectors employed are smaller than those used for revolving lights; and this is necessary, in order that they may be set up on the circular framework with their axes so inclined as to admit of their illuminating every point of the horizon.

REVOLVING APPARATUS ON THE CATOPTRIC PRINCIPLE.

The *revolving* is produced by the rotation of the apparatus; and as it exhibits once a minute, or once in half a minute, a light gradually increasing to a maximum, and then gradually receding into darkness, it has a very impressive effect.

The *revolving red and white* is obtained by the revolution of a framework with red and white lights on different sides or faces: it is susceptible, of course, of considerable variation.

The *flashing* light * is produced in the same manner as the *revolving;* but, owing to a difference in the construction of the

---

* This description of light was first introduced by the late Mr. Robert Stevenson, in 1822, at Buchanness, on the coast of Aberdeen. Mr. Stevenson was also the author of the "intermittent light."

framework, the reflectors are arranged with their rims or faces in one vertical plane, and their axes in a line inclined to the perpendicular; a disposition of the mirrors which, combined with greater quickness of revolution, showing a flash once every five seconds, creates an effect wholly different from that of a revolving light, and presents the appearance of an alternately rising and sinking illumination. As this is but of momentary duration, the light resembles a rapid succession of bright flashes—whence its name.

The *intermittent* light is distinguished by bursting suddenly into view, and continuing steady for a very brief interval, after which it is suddenly eclipsed for an equally brief period. This is accomplished by the vertical motion of opaque cylinders in front of the reflectors—the light being thus alternately revealed and hidden.

The *double* lights ("which are seldom used except there exists a necessity for a *leading line*, as a guide for taking some channel or avoiding some danger") are sometimes exhibited in two towers, sometimes in the same tower at different elevations.

The other distinctions indicated are simply variations of those which we have just described.

As the catoptric is the *reflecting* system, and is distinguished by its use of reflectors, so the dioptric is the *refracting* system, and distinguished by its use of lenses. The former bends the beams of light in the required direction by the agency of reflectors, each with its separate lamp or burner; the latter employs lenses for the same purpose, and arranges them round a fixed central light.

The application of lenses to lighthouse illumination seems to have been proposed in England as early as 1762, but owing to mechanical defects they were found to yield a light inferior to that of the paraboloidal re-flectors, and failed to secure adoption. It was suggested by Buffon that a lens might be constructed in concentric zones out of a solid piece of glass, but the difficulties of the process proved to be insuperable. About 1773 Condorcet proposed that burning lenses should be built up, as it were, in separate pieces; and a similar method was described by Sir David Brewster

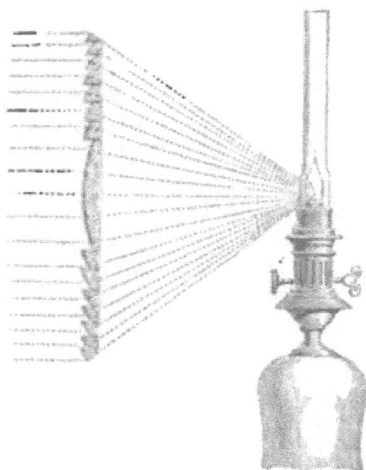

ANNULAR BUILT LENS.

in 1811. Quite independently, Augustin Fresnel arrived at the same result in 1819; and having constructed a suitable lens, he placed a strong light in its focus, and obtained a suitable illuminant for lighthouses. In conjunction with Arago and Mathieu, he introduced a large lamp having four concentric wicks; and, arranging four or more annular lenses in stages around this lamp, he became the inventor of the dioptric system.

Let us visit the lantern of a lighthouse which has been constructed upon this system. Necessarily, the first object to be noticed is the lamp. This is a fixture in the centre of the lantern, and is supplied with oil either on the hydrostatic principle, or, as is usually the case, by the employment of pressure. In the former instance, a cistern is placed near the lamp, but at a higher level than the burner, and by the natural action of gravitation the oil penetrates to the burning-point. In the latter, the oil is forced or pumped up to the wicks from a reservoir underneath the burner by a weighted plunger, adjusted to the requisite amount of pressure. The wicks are arranged concentrically—that is, ring within ring—and they vary in number according to the intensity of light desired.

The next object of interest is the lenticular apparatus —the arrangement of the lenses or prisms, which bend or refract the rays from the central flame, so as to gather them into a compact beam.

A ray of light, as everybody knows, when transmitted obliquely from a transparent body to another of different density, undergoes, at the point where it strikes the common surface of the two planes, a sudden change of direction. This change is called *refraction.* For instance : plunge one-half of a straight ruler into a basin of water. The ruler no longer appears *straight,* but *bent back* or *broken* at the point where it enters the water. This refraction is the great thing to be attained in lighthouse illumination.

Fresnel began by fitting together four or more annular lenses, so as to form a crystal frame which surrounded the lamp. When this frame was made to revolve, the mariner received the full effect of each lens as soon as its axis was directed towards him; but in all other positions there was darkness. In order to utilize the rays of light which passed above the lens, Fresnel employed a system of double optical agents. These consisted of inclined hanging lenses (*a*), which refracted the upper rays, these being reflected into the proper direction by mirrors (*b*) placed above at a proper angle. The lower rays were intercepted by fixed prisms (*pp*),

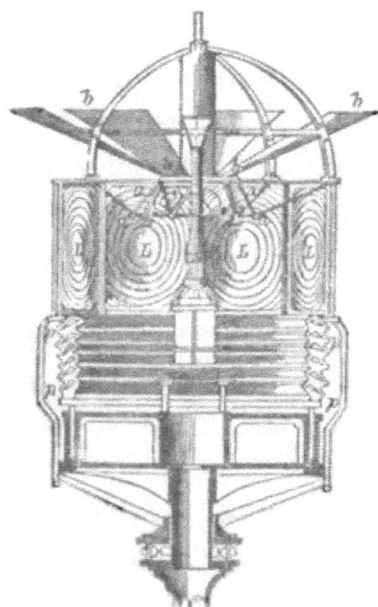

FRESNEL'S REVOLVING LIGHT.

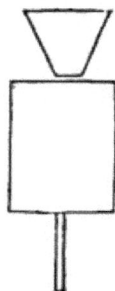

which did not revolve, and, exhibiting everywhere a fixed light, were necessarily of very inferior power to the solid beams proceeding from the annular lenses (*L*), and the arrangement of upper and lower lenses and mirrors.

So far as a "fixed light" was concerned, Fresnel's

apparatus was practically perfect; it illuminated the whole of the horizon with equal intensity. But it was otherwise with revolving lights, in which the whole of the rays have to be collected into one or more beams of parallel rays. As we have seen, those rays which escaped past the lens were caught up by two agents; but both caused loss of light by absorption. Fresnel himself estimated the loss at one-half of the whole incident rays. To remedy this serious defect, Mr. Thomas Stevenson, in 1849, proposed an arrangement, which he called the *holophotal*, for dispensing with one of those agents, and advantageously employing total reflection for revolving lights, as Fresnel had

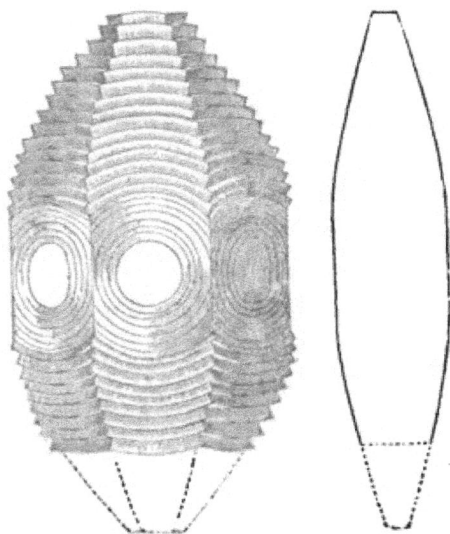

STEVENSON'S HOLOPHOTAL
REVOLVING LIGHT.

used it for fixed lights. His apparatus, as commonly employed, consists of a polygonal framework which circulates round the lamp, each face producing a beam of parallel rays. When the frame revolves with uniform rapidity round the central burner, the mariner is alternately illuminated and thrown into darkness, according as the axis of the holophotes is pointed from or towards him. The difference between Fresnel's revolving light and Stevenson's holophotal, briefly speaking, is this: in the latter one agent does the work which in the former is done by two agents—the *entire* glass frame, with its lenses and reflecting prisms, revolving round the central lamp.

The *dioptric holophotal* system, or that which makes total reflection a portion of the revolving apparatus, was first employed on a small scale, in 1850, at the Horsburgh Lighthouse, near Singapore, and on a large scale, in 1851, at North Ronaldshay, in the Orkneys. It is now in use over all Europe and America.

CHAPTER V.

WE now invite the reader to accompany us on an imaginary voyage round the coast of Great Britain, so that we may take stock, as it were, of the lighthouses and lightships by which it is protected. It will be found, as, indeed, a glance at the chart attached to this volume clearly reveals, that, owing to their number, and admirably chosen position, and range of illuminating power, they surround our island with an unbroken belt of light which warns the navigator of rock and reef and shoal, and points out the way to safe and secure channels. There is absolutely not a mile of our shores undefended by these sea-marks and beacon-lights; or, at least, it may confidently be asserted that if any unguarded places still exist, it will not be long before they are brought into the general system.

Starting from the extreme south-western point of England, we direct our attention, in the first place, to the compact little archipelago of islets and rocks known as the Scilly Islands, the *Cassiterides* of the ancient geographers. So stormy is the climate amongst these masses and fragments of granite, that it is said they do not, on an average, enjoy more than six really calm days in the year. Six of them—St. Mary's, Tresco, St. Martin's, St. Agnes, Bryher, and Samson—are inhabited. On the summit of the island of *St. Agnes* a lighthouse has been established since 1680. A coal-fire was the first illuminant, but oil was substituted in 1790. The lighthouse is a plain, circular building, measuring seventy-four feet in height from base to vane, and exhibiting a white light, revolving every half minute, which in fair weather is visible for seventeen miles. This light, however, was extinguished in 1888, when a new lighthouse was erected on *Giants' Castle Head*, in St. Mary's Isle. On *Round Island*, to the north, a light-tower has been established, which shows a powerful red flash every half minute.

The desolate island of Annette, in this group, is thickly beset with rocks and reefs, and associated with many a sad memory of suffering and death. They have been called "the dogs" of Scilly, and compared to those which, in the ancient fable, hovered round the monster of the Sicilian sea—the terrible Scylla, which, from her cavernous lair, darted forth "her ravening mouth" and dragged down the helpless vessel to destruction.

It was here, on the Gilstone Rock, that Sir Cloudesley Shovel, the gallant old admiral of Queen Anne's reign, was lost, his flag-ship the *Association*, and a crew of eight hundred men, going instantly to the bottom.

The *Eagle,* the **Romney,** and the *Firebrand* also perished (October 22, **1707**). On Jacky's Rock the *Thames* steamer went to pieces (January **4**, 1841), and out of sixty-five sailors **and** passengers only three escaped. On the **Crebawethan** the *Douro* was wrecked (1843), and every person on board drowned. The large steamer *Delaware* was driven ashore here, and forty-five **were** lost out of a complement of forty-seven (December **20, 1871**). And, in a thick fog, the *Hamburg* mail-steamer went ashore in this **perilous** neighbourhood (**May 7, 1875**), with the terrible loss of three **hundred and thirty-one** lives.

The **south-westernmost of these rocks is called the** BISHOP (lat. 49° 52′ 30″), **and here** the **Trinity** House resolved to erect a lighthouse in 1852. The design was supplied by **Mr. James** Walker, and **carried out under** the direction of **Mr.** Nicholas, father of the eminent engineer **Sir** James Douglass. The **cost, including** dwellings on mainland, was £36,559, **18s. 9d.** The **tower is of** gray granite, circular in shape, entirely solid to forty-five **feet** above high **water,** and is one **hundred and** forty-seven feet in height. The **foundation-stone** was laid on July 16th, **1852, and** the **stone-work completed** on August **28th, 1857.** The **light, a double** flashing dioptric light **of** the first order—showing two successive flashes of about **four seconds,** with an eclipse of four seconds between each, and forty-eight seconds **of darkness** after **the second** flash—is visible for sixteen miles, and was first exhibited on September 1st,

1858.* The commissioners who visited the lighthouse in 1859 describe it as a magnificent structure, but the most exposed, perhaps, in the world. The rolling seas send their spray over the top of the lantern. To strengthen the tower against wind and wave, it has

BISHOP ROCK LIGHTHOUSE.

been cased in a new coat of granite, three and a half feet thick, and raised thirty-five feet higher. An explosive fog signal, with one report every five minutes, was placed here in 1887.

* Height of lantern 28 feet, and diameter 14 feet. Cost £1,452, 1s

In lat. 50° 3′ 50″, about ten miles from the Bishop, has been moored the **Seven** *Stones* lightship since 1841. It shows a white light, with three flashes in quick succession, followed by thirty-six seconds' eclipse, visible eleven miles. A fog-siren, during thick or foggy weather, utters three blasts very rapidly every two minutes—a low, a high, and then a low note. This vessel has a crew of eleven men. She rides in forty fathoms, east of the rocks, and has occasionally nearly three hundred fathoms of an inch-and-a-half chain out. "Though this is the most exposed vessel on the coast, the master considers that from the length of the sea she rides easier than vessels moored in a shorter sea. As he expressed it, 'She is always ready for it;' but still her decks are occasionally swept by the waves, and when it strikes her forward, 'it is like a four-pounder going off.'"

In a graphic passage, Mr. Walter White describes the south-western extremity of England, the Land's End, which we now approach. "Those who expect," he says, "to see a towering or far-stretching promontory will be disappointed. We form our ideas from ordinary maps, and imagine England's utmost cape to be a narrow tongue thrust out from the firm shore, along which we may walk to meet the advancing waves. But we find the reality to be merely a protruding shoulder or buttress of the vast irregular bluff that terminates the county. Cape Cornwall, which looks so grand, about two miles **distant,** appears to extend farther to the west than the Land's End.

"Sit still and gaze : the scene grows upon **you.** Here the two Channels commingle with the ocean ; **and far** out as eye can reach, and round on either hand till it **meets** the remotest point of the rugged shore, stretches **the** watery expanse. The billows come tumbling in **and break in** thunder at the **base of the** cliffs, dashing **the impatient** spray well-nigh **to** their summit. You **may descend by** steep paths **to a** lower level, and see **the** cavernous opening **which their** plunging assaults have worn through from one side **of** the buttress to the **other. With** what fury they rush into the recess, and **make horrid whirlpools** behind the **mass which some** day **will be an isolated** member of **the rocky** group scattered along the **shore!** There, on the largest **of the cluster, nearly two miles from** shore, stands the Longships Lighthouse, and **all** between is foam and swirl ; waves running together, and leaping high with the shock : **a** dangerous **channel known as** the Kettle's Bottom. See how the **water** chafes around the Armed **Knight** there on the left, and the Irish Lady on the right, and all the nameless lumps! Yonder, under the **cape, at** the extremity **of** Whitesand Bay, are the Brisons, invested by shipwreck with a fearful interest."

The LONGSHIPS LIGHTHOUSE **was erected in** 1795 by **a Mr. Smith, who received as** his reward the right to levy a toll upon shipping for a limited number of years. It was afterwards purchased of his representatives **by** the Trinity House, and a new tower erected in 1843, which was rebuilt in 1883. The material used is granite, and the large blocks are trenailed on the plan introduced by Smeaton into his great work at the Eddystone. Its circumference at the base measures sixty-two feet. From the base to the lantern vane the height is one hundred and seventeen feet; from high-water mark to centre of lantern, one hundred and ten feet. Yet in heavy weather the waves fling their foam with a rattle against the lantern panes, and on one occasion even lifted the cowl off the top, so that the water poured in and extinguished some of the lamps. At this lighthouse, in foggy weather, is used the new explosive fog-signal. The explosive charge, with a detonator (an enlarged percussion cap filled with fulminate) inside, is connected by the necessary wires with a small electric machine, and suspended at the end of a long arm projecting from the building. On a current being generated by the machine, the induced spark fires the detonator and explodes the charge. There are two reports every two minutes. The light is white, with red sectors, eclipsed for three seconds every minute.

Under the lighthouse, at the end of a long fissure in the rock, is a cavern; and when the billows are very high, the noise produced by the rush and roar of pent-up air from this cavern is so great that the keepers can hardly sleep. It is recorded that one man—we suppose, a new-comer—was so terrified on one occasion that his hair turned suddenly white!

About eight or nine miles to the south-west lies a dangerous mass of greenstone, called the *Wolf's Crag* or *Wolf's Rock*, in the midst of an incessant swirl and eddy of waters. At times the full force and fury of the Atlantic beats upon it. Some years ago an attempt

WOLF'S CRAG LIGHTHOUSE.

was made to erect upon its summit the figure of an enormous wolf in copper, made hollow within, and so constructed that the mouth, on receiving the blasts of the gale, should emit a loud hoarse sound to warn off approaching vessels. The scheme was rendered abortive, however, by the violence of the elements; but a beacon was afterwards set up and maintained here until the erection of a lighthouse could no longer be delayed. In 1862 the building was begun, from the designs of Mr. James Walker; and in 1869 a noble tower of granite, one hundred and ten feet high, rose complete and stately above the waves. In its construction, which cost £62,726, about three thousand two hundred and ninety-six tons of granite were made use of. The tower exhibits one powerful light, white and red alternately, revolving at intervals of half a minute, and visible for sixteen miles.

LIZARD POINT LIGHTHOUSE.

LIZARD POINT is the southernmost promontory of England, and at the entrance to the Channel mounts guard with two large and massive lighthouses, situated east and west, at a distance of two hundred and twenty-

two **feet** apart. Each **tower is of** brick, **sixty-one feet**
high, and perched **on a** mass of serpentine one hundred
and sixty-eight feet above the sea. Between the two—
one of which was erected in 1752, and the other rebuilt
in 1878—are situated the keepers' dwellings and offices.
Coal-fires were burned here down to January 16th, 1812,
when oil was substituted; but the electric light, on **Dr.**
Siemen's system, was introduced on the 29th of March
1878. It is visible for twenty-one miles. This is also
a fog-siren station.

Continuing **our course** along the south coast, we
come in sight **of** *St. Anthony's Point*, Falmouth, lat.
50° 8′ 39″, where an octagonal **tower of** gray granite,
sixty-three **feet high, was erected in** 1835 from **Mr.**
James Walker's designs. **It shows an** upper light,
white, revolving every **twenty seconds; and a** lower
light, white, fixed, **which is exhibited to** clear **the**
Manacle rocks. . Both **have a fair-weather range of**
thirteen **miles.**

We now see before **us the most celebrated of** British
lights—

The EDDYSTONE, **on** the Eddystone rock, **lat. 50° 10′**
49″, long. 4° 15′ 53″, about fourteen miles **south-west**
of Plymouth.

The first lighthouse of **a** regular character erected on
our English coast seems **to** have been that of Lowestoft,
in 1609. Among its successors were those of Hunstanton **Point,** 1665, and **the** Scilly Islands, 1680.   To

the same period belong the lights at Dungeness, Orfordness, and the Eddystone—the last-named being the
most important, the most remarkable, and the most
interesting. All these were at first coal-fires.

1. The reef of rocks, of which the highest summit
is called the Eddystone, being in a line with the
Lizard in Cornwall, and **Start Point in** Devonshire,
lies across the track not only of vessels bound for the
great seaport of Plymouth, but of vessels beating up
and down the Channel. At high water its craggy tops
are hardly visible, and their position is discernible only
from the waves which "eddy" and seethe above them;
at low water, several broken and irregular ridges of
gneiss are exposed. When a gale blows up from the
south-west, they become the centre of a chaos of heaving and hurtling billows, in which no ship could hope
to escape destruction. To avoid them, therefore, vessels were accustomed to enter the Channel in a much
more southerly latitude than is now the case; but in
doing so they frequently exposed themselves to dangers
not less formidable, and hence the numerous wrecks
which occurred on the French coast, and upon the
rocks surrounding the islands of Jersey, Guernsey, and
Alderney. It was a matter, therefore, of national interest when, in 1696, a private gentleman, Mr. Henry
Winstanley of Littlebury, in Essex, resolved to plant
a lighthouse on the Eddystone, and, having obtained
the necessary Parliamentary powers, plunged into his
philanthropic enterprise with characteristic energy.

He was a man with an inventive turn of mind, but hitherto had employed his ingenuity chiefly in the preparation of elaborate practical jokes. If a guest at his house, on retiring to his bedroom, removed an old slipper which was lying in his way, immediately a ghost started up before him. Or if he threw himself into an easy-chair, the two arms suddenly closed in and held him prisoner. Or if, in the garden, he sat down on an arbour seat, he was swiftly set afloat in the middle of the neighbouring fish-pond.

Mr. Winstanley's innate eccentricity was conspicuous enough in the fantastic design of the *first* Eddystone Lighthouse, which had all sorts of projections and quaint contrivances, as if intended to catch every wind that blew. Its erection was begun in 1696. In the first summer—and of course the work could be carried on only during the summer months—twelve holes were excavated in the rock, and iron fastenings secured in them, to receive and hold the substructure. The progress was very slow, for, though it was summer, the weather, according to Winstanley, proved at times of exceeding violence, and for ten or fourteen days consecutively the sea, agitated by "out-winds and the in-rush of the ground-swell from the main ocean," raged and leaped upwards with such fury as completely to bury the works and prevent all approach to them. The second summer was spent in the construction of a solid circular pillar twelve feet in height and fourteen feet in diameter. In the third year the base of the

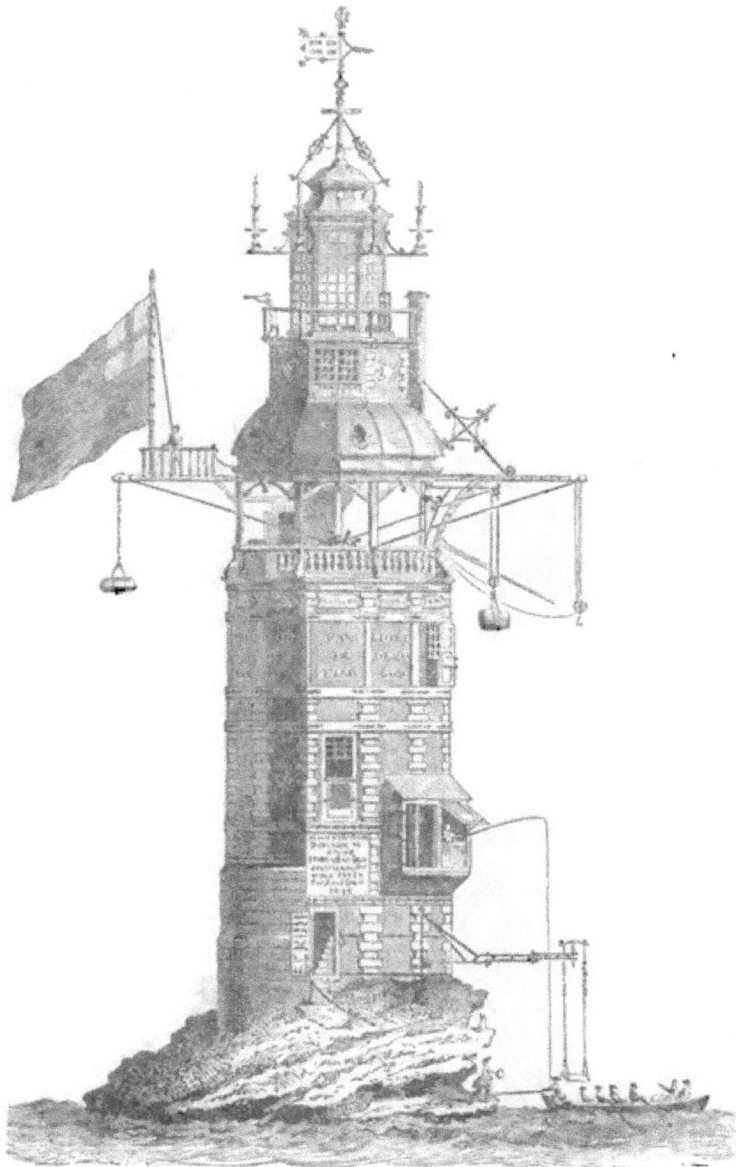

WINSTANLEY'S LIGHTHOUSE AT THE EDDYSTONE.

pillar **was** enlarged two feet, **and the** edifice raised **to a height** of sixty **feet.*** "Being **all** finished," says **Win-**stanley, "with the lantern, **and all** the rooms that were in it, we ventured to lodge there soon after midsummer, for the greater despatch of the work. But the first night the weather came bad, and so continued that **it was** eleven days before any boats could come near us again; and not being acquainted with the height of the sea's rising, we were almost drowned with wet, and our provisions in as bad a condition, though we worked night and day as much as possible to **make shelter for** ourselves. In this storm we lost some **of our** materials, although we did **what we could to** save them; **but** the boat **then** returning, **we all** left **the** house **to be re-**freshed on shore. **And as** soon as the weather did per-mit we returned and finished all, and put up the light [a coal-fire] on the 14th November 1698; which being so late in the year, **it was** three days before Christmas before **we** had relief **to go on shore** again, and were almost **at the last** extremity **for** want of provisions; **but, by** good Providence, then **two** boats came with provisions and the family that was to take **care of the** light; and so ended **this year's work."**

The fourth and **last year was** devoted **to** strengthen-ing the foundations **and** enlarging the accommodation **of** the lighthouse, **which, when** completed, resembled

---

* See Smeaton's "**Narrative of** the Building **of** the **Eddystone Light-**house" (ed. 1791); "Smeaton **and Lighthouses**" (ed. 1844); Dr. Smiles's "Lives of the Engineers."

nothing so much as "a Chinese pagoda." The gallery around the lantern was so wide and open that it was possible, when the sea ran high, for a six-oared boat to have been lifted by the waves and driven through it. If Mr. Winstanley had possessed a practical knowledge of the profession into which he had obtruded himself with so much self-confidence, he would have known that the quaint fabric he had raised was singularly ill adapted to bear the pressure of wind and wave that would inevitably be brought to bear upon it; but he was so proud of his work, and so convinced of its solidity, that he expressed a wish to be beneath its roof in the greatest storm that ever blew under the face of heaven, convinced that it would not shake one joist or beam. His presumptuous egotism was severely punished. With his workmen and keepers he had taken up his residence in the lighthouse, when a terrible gale arose, and on the 26th of November developed an almost exceptional force. All through the night raged the destroying angel; and when morning came the crowds that flocked to the shore at Plymouth discovered that the lighthouse on the Eddystone was gone! No trace was ever found of the unfortunate occupants.

2. Whatever may be our opinion of the temerity and unwise self-confidence of Winstanley, we must needs admit that he at least demonstrated the feasibility of the erection of a lighthouse on the Eddystone rock, and that his achievement deserves to be recorded as

one of the most generous enterprises which any phil-
anthropist has ever undertaken.

After an interval of three years the task of building
a new lighthouse was taken up by a Captain Lovet,
who obtained a ninety-nine years' lease from the
Trinity Corporation, and engaged as architect and engineer a Ludgate Hill silk-mercer, named John Rudyerd. The reasons which determined Lovet in his
apparently curious choice are unknown; at all events,
it turned out sufficiently fortunate. Rudyerd laid
down an elegant and effective design : instead of a
polygon, he chose a circle for the outline; and instead
of the projections and ornaments with which Winstanley had encumbered his building, he aimed at the
most absolute simplicity of construction. In preparing
his foundation, he showed considerable ingenuity. He
divided the irregular area of the rock into seven rather
unequal stages, in which he cut six-and-thirty holes to
the depth of from twenty to thirty inches. These
holes measured six inches square at the top, gradually
narrowed to five inches, and again expanded to nine
inches by three at the bottom. Into these were keyed
stout bolts or branches of iron, each bolt being exactly
fitted in size to the hole it was intended for, and according to its dimensions and form weighing from two to
five hundredweight. These bolts having been securely
fastened, Rudyerd proceeded to lay a course of squared
oak timbers lengthwise on the lowest stage, so as to
reach the level of the stage next above. Another set

of timbers was laid diagonally, so as to cover the first course and raise the level to the surface of the third stage. Alternate diagonal and longitudinal courses were thus placed until a foundation of solid wood had been obtained, two courses higher than the topmost point of the rock, all being firmly fitted to each other and to the rock by the aforesaid iron bolts, and all being exactly trenailed where they intersected one another.

Rudyerd's lighthouse we have frequently seen described as "a wooden structure;" but, in reality, it consisted of a combination of timber and masonry, in the proportion of two courses of timber to five courses of granite, so far as the basement was concerned. It is true that upon this basement or pedestal, which measured twenty-three feet in height, a column of timber was raised for a further height of sixty-nine feet, giving a total of ninety-two feet. The interior was divided into four rooms or stories, above the uppermost of which rose the lantern, an octagon of ten and a half feet, crowned with a ball of two and a fourth feet in diameter. The building was completely finished in 1709.

A couple of anecdotes may be told in connection with Mr. Rudyerd's lighthouse. England and France were at war while it was in course of construction; and a French privateer sent an armed boat's crew to make a descent upon the rock and carry off the workmen. As soon as their capture was made known to Louis XIV., he ordered their immediate release, and that they should be sent back to their work, with some presents to compensate them for the loss and inconvenience they had sustained. "Though at war with England," said the king, "I am not at war with mankind."

Some visitors to the lighthouse, after inspecting its internal arrangements, remarked to one of the keepers that they thought it quite possible for them to live very comfortably in their enforced seclusion. "That might be," said the man, "if we had but the use of our tongues; but it is now fully a month since my partner and I have spoken to each other."

Rudyerd's lighthouse did not have the good fortune to celebrate its jubilee. On the 2nd of December 1755, through some cause unknown, it caught fire. Three keepers were on the rock at the time; and when one of them, whose turn it was to watch, entered the lantern, at about two in the morning, to snuff the candles, he discovered that it was filled with smoke, and on his opening the door which led to the balcony flames immediately burst forth. He at once alarmed his companions, and they used every exertion to extinguish the fire; but owing to the difficulty of raising a sufficient supply of water, and the dryness of the woodwork, its ravages could not be checked, and they were compelled to retreat downwards from story to story. Happily, at dawn the burning lighthouse was descried by some fishermen, who hastened with the news to shore, and a well-manned boat was instantly put off to the rock. It reached the Eddystone at ten o'clock, when the fire had been burning for eight hours, and the keepers, half-dazed with terror, were found in a hollow on the east side. No sooner were they landed at Plymouth than one of them, strange to say, immediately made off and was never heard of again. So singular a circumstance gave rise to the suspicion that he had originated the fire; but when we remember that a lighthouse supplies its occupants with no means of retreat, and that the probability is they will perish with it, we can hardly believe it to be the place an incendiary would choose for the practice of arson. The man's flight seems to us more easily and reasonably explained by the demoralizing influence of panic-terror; he had lost control over himself, and fled precipitately in the hope of escaping from his fears.

Henry Hall, one of the other two keepers, met with a remarkable death. While engaged in throwing water against the burning roof of the cupola, he happened to look upwards, and a quantity of molten lead dropped on his head, face, and shoulders, scalding him most severely. His mouth was open at the time, and he asserted that a portion of the lead had gone down his throat. The physicians who attended him regarded his story as incredible; but the man continued to grow worse, and on the twelfth day of his illness, after some violent spasms, expired. A *post-mortem* examination revealed the truth of his statement, for in the stomach was found a flat oval lump of lead weighing about nine ounces.

In the course of its brief history, Rudyerd's lighthouse was the scene of a tragical episode. For some years after its establishment, it was in the charge of two keepers only, who were on duty four hours alternately. Each, when his watch ended, was bound to call the other, and see him installed in the lantern before he himself quitted it. On one occasion, when the keeper going off duty went to call his colleague, he found him dead. At once he hoisted his flag on the balcony, whence it was visible at the Rame Head, near Plymouth, and waited eagerly for the assistance it was understood to summon. Unfortunately, the weather became so violent that no boat could put off from the shore, and the lonely keeper was left to the miserable companionship of a dead body. It is not easy to imagine a more painful situation, as the poor man was afraid to fling the corpse into the waves, lest he should be charged with the murder of his companion, and yet the stench proceeding from it threatened him with serious illness. At length he contrived to drag it up to the balcony, and to fasten it there. It was nearly a month before relief arrived, and then the dead body was found in such a condition that it could not be removed to Plymouth for interment, but without further delay was consigned to the deep. Thenceforward three keepers were always employed, so that in case one should fall ill or die, two might still be on duty.

3. The great usefulness of a lighthouse upon the Eddystone had by this time been so abundantly de-

monstrated that the proprietors took immediate steps to replace **Mr.** Rudyerd's building. **For this purpose, at** the instigation of Mr. Robert Weston, one of **their** number, they called into requisition the services of **Mr.** John Smeaton, a mathematical instrument-maker **by** profession, but a man of remarkable capacity, who had already given proofs **of** mechanical inventiveness and **resource.** **He** was, in truth, a mechanic **born :** in his **childhood** mechanical tools were his playthings, and **before his** sixth year **he had** designed a windmill and the model of a pump. When at school **he** occupied his holidays with mechanical pursuits. At the age of fourteen he was an adept with the smith's hammer and the turner's lathe. **He forged** his iron and steel **and** melted his metal. These tastes still clung to him when he became a mathematical instrument-maker. He invented a machine to measure a ship's way at sea, as also a compass of peculiar construction, besides submitting to the Royal Society some improvements which he had contrived in the air-pump, and experiments on the natural power of water and wind to turn mills and other machines dependent upon circular motion. But **the true** extent of his remarkable abilities might never have been known **if** he had not been intrusted, at the age of thirty-two, with the erection of the *third* lighthouse upon the Eddystone.

On examining into the nature of the work before him, **he came to** the conclusion that **both Winstanley** and Rudyerd's lighthouses **had been deficient in** *weight* and *strength;* and he resolved to build a structure of such solidity that the sea should give way to the light-house, not the lighthouse to the sea. He determined, therefore, that the building should be entirely of stone; and to secure stability, he proposed, while retaining the idea of a cone, to enlarge the diameter of the base, and, generally speaking, to take as his model "the natural figure of the waist or bole of a large spreading oak."[*] "Connected with its roots," he says, "which lie hid below ground, it rises from the surface thereof with a large swelling base, which at the height of one diameter is generally reduced by an elegant curve, con-cave to the eye, to a diameter less by at least one-third, and sometimes to half, of its original base. From thence, its taper diminishing more slowly, its sides by degrees come into a perpendicular, and for some height form a cylinder. After that a preparation of more circum-ference becomes necessary, for the strong insertion and establishment of the principal boughs, which produces a swelling of its diameter. Now we can hardly doubt but every section of the tree is nearly of an equal strength in proportion to what it has to resist; and were we to lop off its principal boughs, and expose it in that state to a rapid current of water, we should find it as much capable of resisting the action of the heavier fluid when divested of the greatest part of its clothing, as it was that of the lighter when all its

---

[*] This is Smeaton's own statement, though criticised sharply by Mr. Alan Stevenson in his "Treatises on Lighthouses," p. 98.

spreading ornaments were exposed to the fury of the wind; and hence we may derive an idea of what the proper shape of a column of the *greatest stability* ought to be, to resist the action of external violence, when the *quantity of matter* is given whereof it is to be composed."

The first work done on the rock was in August 1756; but the autumn months were mainly occupied in the transport and preparation of the granite and other materials, and in excavating the steps or stages for the reception of the foundation.

Early in June 1757, Mr. Smeaton resumed operations. On the 12th, the first stone, weighing two tons and a quarter, and inscribed with the date in deep characters, was set in its place. Next day the first course was finished, consisting of four stones, so ingeniously dovetailed together and into the rock as to prove a compact mass, from which it was impossible to separate any particular stone. The sloping form of the rock rendered only this small number of stones necessary for the first course; but the diameter of each course increased until the top of the rock was reached—so that the second course, laid on June 30th, consisted of thirteen stones; the third, on July 12th, of twenty-five; the fourth, on July 31st, of thirty-three. The sixth course was laid on August 11th, and rose above the general wash of the tide; and Smeaton was able to congratulate himself on having completed the initial stage of his enterprise.

Up to this level, in every course, the stones had not
only been securely dovetailed into the rock, but also
made fast by oaken wedges and cement. To receive
these wedges, a couple of grooves were cut in the waist
of each stone, from the top to the bottom of the course,
an inch deep and three inches wide. We borrow from
Smeaton's own narrative his description of the manner
in which each stone was laid :—

"The stone to be set being hung in the tackle, and
its bed of mortar spread, was then lowered into its
place, and beaten with a heavy wooden mall, and
levelled with a spirit-level; and the stone being ac-
curately brought to its marks, it was then considered
as set in its place. The business now was to retain it
exactly in that position, notwithstanding the utmost
violence of the sea might come upon it before the
mortar was hard enough to resist it. The carpenter
now dropped into each groove two of the oaken
wedges, one upon its head, the other with its point
downwards, so that the two wedges in each groove
would lie heads and points. With a bar of iron about
two inches and a half broad, a quarter of an inch thick,
and two feet and a half long, the ends being square,
he could easily (as with a rammer) drive down one
wedge upon the other; very gently at first, so that the
opposite pairs of wedges, being equally tightened, they
would equally resist each other, and the stone would
therefore keep place. A couple of wedges were also,
in like manner, pitched at the top of each groove; the

dormant wedge, or that with the point upward, being held in the hand, while the drift-wedge, or that with its point downward, was driven with a hammer. The whole of what remained above the upper surface of the stone was then cut off with a saw or chisel; and, generally, a couple of thin wedges were driven very moderately at the butt-end of the stone; whose tendency being to force it out of its dovetail, they would, by moderate driving, only tend to preserve the whole mass steady together, in opposition to the violent agitation that might arise from the sea."

The stone having been securely set, a certain quantity of mortar was liquefied; and after the joints had been carefully pointed, this liquid mortar or cement was poured in with iron ladles so as to fill up every void. The more consistent parts of the cement naturally fell to the bottom, while the stone absorbed the watery; the vacancy thus created at the top was repeatedly filled up until the whole was entirely solid; the top was then pointed, and, where necessary, defended by a layer of plaster.

To insure the solidity of the superstructure, some other means, however, was necessary. Accordingly, a hole one foot square was cut right through the middle of the central stone in the sixth course, and at equal distances in the circumference were sunk eight other depressions of one foot square and six inches deep, for the reception of eight cubes of marble, in masonry called *joggles*. A strong *plug* of hard marble, from the rock near Plymouth, one foot square and twenty-two inches long, was set with mortar in the central cavity, and driven firmly into it with wedges. As this course was thirteen inches high, it is evident that the marble plug which reached through it rose nine inches above the surface. Upon this was fixed the central stone of the next course, having a similar bore in the middle, bedded with mortar, and wedged as before. By this means no pressure of the sea acting horizontally upon the central stone, unless it was able to cut in two the marble plug, could move it from its position; and the more effectually to prevent the stone from being lifted, in case its mortar-bed should accidentally be destroyed, it was fixed down by four trenails. The stones surrounding the central were dovetailed to it in the same manner as before; and thus one course rose above another, with no other interruption than the occasional violence of the waves or the inclemency of the weather.

In every stage of the laborious and difficult work Smeaton made his personal energy distinctly felt. When the solid tower had risen high enough to assume the form and appearance of a level platform, he proposed to enjoy the limited promenade which it afforded; but making a false step, and failing to recover himself, he fell over the brink of the work and among the rocks on the west side. The tide having retired, he sustained no very serious injury; but he dislocated his thumb, and as no medical assistance was possible, "laid fast hold of it with the other hand and gave it a violent *pull*, upon which it snapped into its place"— an instance both of the promptitude and resolution of the man.

The ninth course was laid on the 30th of September, and operations were suspended for that year. The following winter proved to be exceedingly tempestuous and protracted, so that it was May 12th before he and his gallant little company of artificers again saw the Eddystone. To their great satisfaction they found the entire work in exactly the same condition as when they left it. The cement had grown as hard as the stone itself, and the whole was concreted into one solid mass. Thenceforward the operations of the builders made rapid progress, and by September the twenty-fourth course was laid, completing "the solid" part of the structure, and forming the floor of the store-room. But as Smeaton fully understood how great would be the gain to navigation if a light could be shown that winter, he resolved on a vigorous effort to finish the store-room and erect a light above it. The building had been carried up solid as high as there was any reason for supposing it would be subjected to the rush of the waters—that is, to thirty-five and a quarter feet above its base, and twenty-seven feet above the top of the rock. At this elevation the diameter was reduced to sixteen feet eight inches; and it became necessary to make the best possible use of this space, consistent

with a due regard to the primary and indispensable condition of *strength.* The rooms, therefore, were restricted
to a diameter of twelve and a quarter feet, leaving
for the walls a thickness of two feet two inches. These
walls were constructed of single blocks of granite, in
such wise that a complete circle was formed by sixteen
blocks, which were cramped together with iron, and
also secured to the underneath course by marble plugs
as before. To prevent the water from penetrating
through the vertical joints, flat stones were inserted
into each, so as to be lodged partly in one stone and
partly in another. With all these ingenious devices,
course the twenty-eighth was completely set on Saturday, September 30th. This and the next course received the vaulted floor—the vaulted floor of the
upper store-room, but also the ceiling of the lower
store-room, for it answered both purposes. Smeaton
then rapidly proceeded with arrangements for a temporary light-room and keepers' accommodation, when,
on October 10th, a prohibition arrived from the Trinity
Corporation based upon certain legal difficulties, and
he terminated operations for the season.

The work was resumed for the fourth and last season
on July 5th, 1759. The second story was finished by
the 21st; the third, by the 29th; and on August 17th
the column or shaft of Smeaton's noble and graceful
structure—upon which all later rock lighthouses of the
first class have more or less directly been modelled—
was completed (forty-six courses of masonry, and a

height of seventy feet). The dimensions were **as follow :—**

	Feet.	In.
The six foundation courses to the **top of the rock**	8	4¾
The eight courses to the entry-door	12	0½
The ten courses of the well-hole to the store-room **floor**	15	2¼
The height of the **four** rooms to the balcony floor	34	4½
Height of the main column	70	0

The lantern was then erected, and on the last stone
set, that over the door of the lantern on the east side,
was engraved the inscription—" 24th August **1759.**
LAUS DEO." On the course under the ceiling of **the**
upper **store-room** Smeaton traced, in grateful **acknow-**
ledgment **of** the protection **of Providence,** the verse
from Psalm **cxxvii.,** "Except **the Lord** build the **house,**
they labour in vain **that build it."** The iron-work of
the balcony and the lantern **were next set up, and** the
whole was surmounted **by a gilt ball.**

The different rooms **or stories of** the lighthouse **were**
thus appropriated :—

First, the store-room, **with** entrance door. **Second,**
the upper **store-room.** Third, the kitchen, **with fire-**
place and **sink, two settles with** lockers, **a** dresser with
drawers, two cupboards, and a rack **for** dishes. Fourth,
the bedroom, with three cabin-beds, **three** drawers, and
two lockers. And fifth, the lantern, **in which a** seat
was placed all round, except at **the door, which opened**
on the balcony or gallery.

In the upper store-room were **inserted two** windows,
and four windows **each in the kitchen and** bedroom.

SMEATON'S LIGHTHOUSE AT THE EDDYSTONE.

In fixing their bars an accident happened to Smeaton, which was nearly attended with fatal results. "After the boat was gone," he says, "and it became so dark that we could not see any longer to pursue our occupations, I ordered a charcoal fire to be made in the upper store-room, in one of the iron pots we used for melting lead, for the purpose of annealing the blank ends of the bars; and they were made red-hot all together in the charcoal. Most of the workmen were set round the fire, and by way of making ourselves comfortable, by securing ourselves and the fire from the wind, the windows were shut, and, as well as I remember, the cover or hatch put over the manhole of the floor of the room where the fire was—the hatch above being left open for the heated vapour to ascend. I remember to have looked into the fire attentively to see that the iron was made hot enough, but not overheated; I also remember I felt my head a very little giddy; but the next thing of which I had any sensation or idea was finding myself upon the floor of the room below, half drowned with water. It seems that, without being further sensible of anything to give me warning, the effluvia of the charcoal so suddenly overcame all sensation, that I dropped down upon the floor; and had not the people hauled me down to the room below, where they did not spare for cold water to throw in my face and upon me, I certainly should have expired upon the spot."

The glazing of the lantern having been completed, a

light was once more shown from the Eddystone rock
on the night of Tuesday, October 16th. According to
our present ideas of lighthouse illumination, its power
was inadequate and unsatisfactory, for it consisted only
of four-and-twenty candles, arranged in a couple of
iron circles, suspended from the roof of the lantern
like chandeliers; but it was a great improvement
upon anything which had previously been attempted.
Oil lamps were substituted about 1810, and in 1845
Fresnel's dioptric system, with a four-wick lamp, yield-
ing a beam of light visible at a distance of thirteen
miles.

For a hundred years Smeaton's lighthouse gallantly
withstood the fury of wind and wave,—a noble monument to the genius of its architect and builder. But,
latterly, reports reached the Trinity House authorities
that the main tower had begun to oscillate and vibrate
to such an extent as greatly to alarm the keepers,
and, accordingly, workmen were at once employed to
strengthen it. The precautionary measures taken
failed, however, to restore the stability of the building;
and in 1877, a close examination revealed the unsus-
pected fact that the source of weakness was not in
Smeaton's tower but in the rock itself, which had become undermined by the action of the waves. It was
found necessary, therefore, to provide at once for the
erection of a new lighthouse on a foundation of a more
permanent character; and into this new structure it
was resolved, of course, to introduce all the improve-

ments which engineering science had devised since the completion of Smeaton's work.

This great undertaking was intrusted to Sir James Douglass, the engineer to the Trinity House Corporation —no unworthy successor, be it said, to John Smeaton. After surveying the spot, he chose as a suitable site for the projected building a rock situated about forty yards from the old Eddystone in a south-south-east direction; and the necessary preparations having been made, the workmen effected their first landing on July 17, 1878. As the foundation had to be laid below the level of low water, a coffer-dam had to be constructed for the protection of the men while working, and the rock to be cut away for the reception of the foundation courses of masonry. By the 21st of December, when operations were suspended for the winter, one-fourth of the dam had been completed, and 1,500 cubic feet of rock excavated. Mr. Edwards describes this period, while the men were at work below low-water mark, as the most perilous.* "Not more than three hours at a time could be spent on the rock by the working party. From about three-quarters ebb to about three-quarters flood tide was the limit of their stay, and during that interval the utmost energy of all had to be exerted. With a rough sea, landing on the rock was simply out of the question; but often when at work, the party

* Edwards, "The Eddystone Lighthouses," ed. 1882, from which, and from articles in the *Times* and other daily papers, the following account is compiled.

having perhaps effected an easy landing, the sea would get up, and then it would be necessary for all to seize their tools and hurry off to the boats as quickly as possible. Delay would probably mean being hauled off through the water, for no boat could venture near the rocks while the seas were breaking upon them. Occasionally with a smooth sea there is a kind of under swell which breaks with great force upon any obstacle interposed in its path. These 'rollers,' as they are called, are supposed to be caused by some disturbance in mid-ocean, and at times three or four will follow each other quickly. The look-out man, or 'crow,' watches for any indications of the sea getting up or of 'rollers' coming along, and shouts a warning to the men. The rollers break completely over the rock, and the men, each wearing a life-belt, have simply to hold on to iron stanchions until they have passed, taking care at the same time that they do not lose their tools."

Operations were resumed on February 24th, 1879. The coffer-dam was finished early in June, and then began the work of laying the masonry. On August the 19th the foundation stone was laid, with due Masonic ceremonies, by H.R.H. the Duke of Edinburgh, Master of the Trinity House, in the presence of the Prince of Wales, and many other " Elder Brethren " of that distinguished corporation. Before the termination of the season, on December 19th, eight courses of the foundation were laid; and next year (1880) the tower

DOUGLASS'S LIGHTHOUSE AT THE EDDYSTONE.

was completed up to the thirty-eighth course. On June
1st, 1881, the Duke of Edinburgh laid the top stone
of the tower, in the construction of which 2,171 stones
had been used, equal to 4,668 tons, or about five times
the weight of Smeaton's building. Sir James Douglass's
tower differs from Smeaton's in other and important
respects. It is planted on a cylindrical base, which
offers advantages for landing from a boat, and supplies
the keepers at low water with a pleasant promenade.
Up to a height of twenty-five and a half feet above
high-water mark the tower is solid, with the exception
of a large water tank let into the solid. Whereas in
Smeaton's tower there were only four living-rooms,
besides the lantern, in the new tower there are nine,
each loftier and more commodious than any of
Smeaton's. All have domed ceilings, with an elevation
of nine feet nine inches from the floor to the apex of
the ceiling. The diameter of the rooms is fourteen feet,
except the lower oil-room, which is twelve feet nine
inches, and the entrance-room eleven feet six inches.

The lowest story or compartment is the *water tank*,
which can contain three thousand five hundred gallons.
The relieving vessel, which visits the lighthouse once a
fortnight, brings a supply of fresh water, with other
stores.

The *entrance story* communicates with the entrance
doors, which are made of gun-metal and weigh one ton.
These are approached from outside by a ladder, formed
by gun-metal rungs inserted into the masonry.

Next we ascend into the *lower* and *upper oil-rooms*, where, in eighteen huge cisterns, two thousand six hundred and sixty tons of rape or colza oil are stored. This quantity will feed the light for about nine months.

On the fifth story we enter the *store-room*, which contains a stock of all the materials requisite for maintaining the lighting apparatus and the establishment generally in perfect cleanliness and good order. A supply of coals (four tons) is kept in a coal-bunker here.

In the *crane-room* is fitted up the crane, which works a winch in the store-room below, this winch being used to hoist up the supplies brought by the relieving vessel. This room is also fitted up with cupboards for the reception of provisions, medicines, and dry stores.

The *living-room* occupies the next story. It is provided with a cooking-range, table, chairs, carpet, curtains, and various comforts and conveniences; and a cupboard is stocked with crockery and culinary utensils on an ample scale.

Then comes the *low-light chamber*, from which a white, fixed, subsidiary light is shown so as to mark some dangerous rocks known as the Hand Deeps. The light is produced by a simple arrangement of two large Argand lamps with reflectors.

In the *bedroom* five berths are fitted up round the room in two tiers, three being reserved for the keepers, and two being held available for officials or workmen who may at any time be detained in the tower. Here also is accommodation for the wardrobes and personal effects of the keepers.

On the story next above we find the *service-room*, or office, where are kept all official books and papers, timepiece, barometer, thermometer, and other apparatus and appliances. A careful record of the weather is kept daily, also of the expenditure of oil and stores, and of any incident which may present itself to the notice of the keeper on duty. Necessarily, the most important part of the keepers' work is done at night. Each of the three keepers is on watch for a period of four hours, during which period he must remain either in the service-room or the lantern. Throughout the interior of the tower passes a central iron column, which is hollow, and contains the weight and chain necessary to set in motion the clock-work machinery for producing and regulating the rotation of the glass lens apparatus around the burners, by which the light exhibits two successive flashes every half-minute. The weight is contained in that part of the column which traverses the two lowest rooms, the fall being fifteen and a half feet altogether. The column passing through the two uppermost rooms holds the machinery for winding up this weight, to do which (and it is no light task, the weight being equal to one ton) is part of the watchman's duty, and has to be performed every hour, or if the bell be sounding, as is always the case in foggy weather, every forty minutes.

Lastly, we arrive at the lantern, "a splendid piece of work," cylindrical in shape, with an elevation of sixteen and a half feet, and a diameter of fourteen feet, constructed by Messrs. Chance of Birmingham. The light is furnished by two concentric six-wick burners, each wick producing a flame with a diameter of four and a half inches, equal in intensity to that of seven hundred and twenty - two standard sperm candles. These burners are superposed one upon the other, with a vertical distance between the two of six and a half feet. With both lamps burning, the combined illuminating power is said to be equal to a quarter of a million of candles, or about six thousand times the strength of Smeaton's light. They are used only in foggy weather, when it is desired to send forth flashes of enormous intensity.

The glass apparatus by which this result is accomplished consists of "a twelve-sided drum, each side, also called a panel, six feet three inches in height and one foot eight inches in width, being formed by a central lens, or, as it may popularly be called, a bull's eye, and surrounded by concentric rings of larger bull's eyes, by which the same effect is obtained as though a portion of one huge lens of great thickness and weight, as large as the whole panel, were employed. For purposes which will presently be apparent, the two bull's eyes of adjoining panels are brought close together, very much as though they were two eyes squinting, so that only lengthways they are in the middle of the panel. On the rotation of this twelve-panelled drum, with the inside central light burning, each bull's eye with its surrounding rings carries round a concentrated beam of light, which becomes visible to the outside observer as soon as by a rotation of the apparatus the focus of the bull's eye falls upon him. Now two bull's eyes are, as has been stated, brought close together—so close indeed that a small portion of each is cut off; consequently a very short interval occurs between the flash of the first and that of the second reaching the observer. Thus it will be seen the two flashes occur in quick succession, and then nearly half a minute elapses before another pair of squinting eyes come round and discharge their two flashes. This description applies to one light only; with the two lamps one over the other, two drums superposed are employed, one for each light, the two being identical in all respects, and arranged so as to coincide exactly with each other. The height of the whole apparatus is consequently twelve and a half feet, and with both lights burning a magnificent effect is obtained."

The new Eddystone is equipped with a couple of large bells, suspended under the lantern gallery, each weighing two tons. In foggy weather these are rung automatically, so as to give two sounds every half minute. Due provision has been made against risk from lightning, a conducting rod being led down the tower and attached to the rock, with its extremity some feet below the level of low water.

The cost of the entire structure was estimated at £78,000.

The next object of interest in our coast-survey is *Start Point*, in lat. 50° 13′ 18″, where, at a height of two hundred and four and one hundred and eighty-one feet above the sea respectively,* two lights are shown, from a white circular tower, erected in 1836. The upper revolves at one minute intervals; the lower, twenty-three feet below the lantern, is a fixed white light, indicating the Skerries bank. A fog-siren is established here, giving three strident blasts in swift succession every three minutes.

Mr. Walter White, in one of his pleasant records of home-travel, graphically describes the Start Point Lighthouse :—" A substantial house, connected with the tall circular tower, in a walled enclosure, all nicely whitened, is the residence of the light-keepers. The buildings stand within a few yards of the verge of the cliff, the wall serving as parapet, from which you look down on the craggy slope outside and the jutting rocks beyond— the outermost point. You may descend by the narrow path, protected also by a low white wall, and stride and scramble from rock to rock with but little risk of slipping, so rough are the surfaces with minute shells.

---

* That is, measuring from high-water mark to centre of lantern. The tower itself is ninety-two feet high, and was built from the designs of the late James Walker, at a cost (including the adjoining buildings) of £5,892, 13s. The lantern is twenty-four and a half feet high, with a diameter of fourteen feet.

"A rude steep stair, chipped in the rock, leads down still lower to a little cove and a narrow strip of beach at the foot of the cliffs. It is the landing-place for the lighthouse-keepers when they go fishing, but can only be used in calm weather.

"The assistant-keeper spoke of the arrival of a visitor as a pleasure in the monotonous life of the establishment. Winter, he said, was a dreary time, not so much on account of cold, as of storms, fogs, and wild weather generally. In easterly gales the fury of the wind would often be such that to walk across the yard was impossible; they had to crawl under shelter of the wall, and the spray flew from one side of the Point to the other. But indoors there was no lack of comfort, for the house was solidly built and conveniently fitted, and the Trinity Board kept a small collection of books circulating from lighthouse to lighthouse."

The narrow pebbly ridge of *Hurst Point*, opposite Yarmouth, in the Isle of Wight, with its block-house or fort of the time of Henry VIII., is well known to the yachtsmen of the Solent. It exhibits a couple of guiding lights in its two towers. The Low Tower, on the beach, fifty-two feet in height, is built of brick coated with cement, and was erected in 1812, from the designs of R. Jupp of London; it shows a fixed white light, with a range of ten miles. The High Tower, two hundred and two yards to the north-east, also designed by Mr. Jupp, dates from 1812, and

measures eighty-seven feet from **base to vane**.　**Like**
the Low Tower, **the** material employed in its **construction** is brick.

On the lofty chalk cliff of *Portland Bill* a coal-fire
was exhibited **as early** as 1716, **for** which **an** oil-lamp
was substituted **in August 1788**.　The two lighthouses
are both circular in shape, **and built of** stone.　They
are five hundred and three yards apart.　The higher,
measuring fifty feet from base **to** vane, shows its light
at an elevation of two hundred and **ten** feet above the
sea ; the lower, fifty-five feet from base to vane, is only
one hundred and thirty-six feet above high-water mark.
In both the illuminating apparatus is first-order dioptric, with fixed white lights, visible twenty-one and
eighteen miles respectively.

We pass the Minquiers lightship **(lat. 50° 35′ 30″),**
and Anvil Point Lighthouse, with a white light flashing
every ten seconds, and **come to the** *Needles Point*, the
bold westernmost extremity **of the Isle of** Wight, a
narrow chalky peninsula, nearly **severed** from the mass
of the island by the small stream of the Yare, which
flows into the Solent **at** Yarmouth.　On the summit
of the glittering cliff, at an **elevation** of four hundred
and seventy-four feet above **the** sea, a light was first
shown on the 29th September **1786.**　Notwithstanding
its great elevation, **we find** it recorded—though the
statement seems to **us** incredible—that its windows
were sometimes broken by stones hurled against them

by the raging billows. It had ten Argand lamps, and the same number of plated reflectors, and its light, in clear weather, could be seen eleven miles off. Seven hundred gallons of oil were annually consumed. On windy nights the blaze attracted numbers of small birds, which dashed themselves against the lantern panes and were killed.

Owing to its exceptional elevation, and the mists and sea-fogs which constantly encompassed it, this lighthouse, which was rebuilt in 1854, proved to be of little service to mariners; and therefore, in 1859, the Trinity House resolved that a new one should be constructed on the outermost of the celebrated chalk rocks, called the Needles, which form the seaward prolongation of the point. Whether the Needles derive their name from their wedge-like shape and acute apex, or, as it has ingeniously been suggested, from the German *nieder fels*, "under cliff," we will not here attempt to determine; but their picturesque conformation is well known to visitors to the Isle of Wight. And seamen are well aware of the danger of their proximity in foggy weather. To obtain a foundation for the new building, the rock was cut away almost to the water's edge; cellars and storehouses were also excavated from the chalk. The new lighthouse, designed by the late James Walker, and erected at a cost of £19,850, 5s., is about one hundred and nine feet in height from the base to the top of the ball. It shows an occulting light every minute, with white, red, and green sectors in succession, the white in the direction of the Solent, and is visible (red) nine and (white) fourteen miles. A fog-bell rings in stormy weather through mechanical

NEEDLES LIGHTHOUSE.

agency; its sounds can be heard at a distance of five miles. The total height of the lantern is twenty-eight and a half feet; diameter, fourteen feet.

But this is not the only lighthouse in the Wight.

Dropping down its southern coast we come in sight of a picturesque green ascent, known as St. Catherine's Hill, which rises seven hundred and sixty-nine feet above the sea-level, and commands the rock-bound sweep of Chale Bay, where shipwrecks have unhappily been numerous. As far back as 1323, one Walter de Godyton erected on the summit of this hill a small chantry, which he dedicated to the patron saint of hills and high places, St. Catherine. He also provided an endowment for a priest to chant masses and keep a light burning throughout the hours of darkness for the assistance of mariners approaching the south coast of the island. These duties were faithfully performed until Henry the Eighth's suppression of the smaller religious houses in 1536, when the priest and his endowment disappeared, though the strong-built walls of the chantry remained, and, some years ago, were carefully repaired on account of their value as a sea-mark. It is octagonal in form, and thirty-five and a half feet in height.

In 1785 a lighthouse was erected on this height, though promptly discontinued when it was found that the mists which so frequently encompass the summit rendered the light of comparatively little service. But the dangerous nature of the coast being emphasized by repeated disasters, the Trinity Board felt it necessary to provide for its better protection; and in 1888 the erection of a lighthouse was begun on *St. Catherine's Point*, which projects from the southern spur of the hill. Its lantern was lighted up for the first time on the 1st of March 1840. Dimensions of the tower:— From the terrace to the **top of** the stone-work, **one** hundred feet. Height of lantern and pedestal, **one and a** half foot. Extension of glass frame, **ten feet.** Roof, **ball,** vane, and **lightning** conductor, **eleven and a** half feet. Total elevation of tower, one hundred and twenty-three feet. It is an exceedingly graceful octagonal structure, with turreted **parapet,** designed by the late James Walker; **and the traveller** who, rounding the base of the hill, **comes suddenly upon it, and** sees its **white** and shapely form springing **up** against the azure sky, is sensible of an impression **of** surprise and pleasure.[*]

The interior **measures fourteen feet** in diameter, and one hundred **and fifty-two steps** lead **up to the** lantern-room. Formerly, the illuminating apparatus consisted of a **lamp three and a half inches in** diameter, with four concentric wicks, reflected through **a lens** surmounted by two hundred and fifty mirrors; **but in 1887 this** was **replaced by the electric arc** light, **which** shows **a** flash every **half minute.** There is also a fog-siren at this station.

Off Bembridge Point, and at the western entrance to the Solent, is moored the **Nab** lightship (lat. 50° 42′ 15″), carrying a white light, with a double flash every forty-five seconds. From this **post** can be heard the gong of the *Warner* lightship (lat. 50° 43′ 40″), farther

---

[*] Cost of lighthouse and adjoining buildings, £7,673, 17s. 2d.

up the Solent, **and** the light of the *Owers* **is** visible **in clear weather.**

The *Owers* **lightship** lies in lat. 50° **38′ 35″**; its **record dates from 1788.** It is distinguished from neighbouring sea-marks **by** its revolving light, with its half-minute **flash; red and** white alternately. When **the** commissioners visited **it** in 1858, the master stated **that his father** had **served on board** it exactly **half a** century, and he himself **for forty-two years.** How strange **an existence! and surely, at times, not altogether a** pleasant **one, seeing that, in bad weather,** the vessel occasionally **rode so heavily that** the master **could not lie** on **the floor of** his cabin "without holding on to the **legs of the table."**

On *Beachy Head*, **the seaward** termination **of the Sussex** Downs,—memorable **in our naval history in** connection with the indecisive engagement, on June 29, **1690, between** Admiral Torrington's fleet **and the French,** fought **in the** neighbouring **waters,—or, more** strictly speaking, on *Belle* **Tout, the second cliff west-ward of the great** high promontory, than **which it is three hundred feet lower, is planted a circular white** tower of stone, with a facing of granite, designed **by** James Walker, forty-seven feet high. The light is **two hundred and** eighty-five feet above **the sea, and the apparatus** displays **a flashing light every fifteen seconds, visible over three-and-twenty** miles. **As early as 1670 a light was exhibited from this station.**

Passing the *Royal Sovereign* lightship, on **the Royal Sovereign shoals, in** lat. 50° **41′ 40″, with a light which** shows three quick flashes every forty-five seconds,[*] **and** the lights of Eastbourne, Hastings, and Rye, **we come** to *Dungeness*, the southernmost point of **Kent, where** there are two **lighthouses. The principal one, a circular** tower, ninety-**two feet high**, is painted with horizontal bands of **red and** white. **It was** built by Wyatt **in 1792, after the model of Smeaton's** Eddystone, and at **the expense of the Earl of Leicester, in** the place **of a** quaint old building erected in the reign of **James the** First by a goldsmith named Allen. **It shows a fixed** white light, with **red sector**, visible fifteen miles. **The second station**, dating from **1875, is on the very edge of the** shore, **about a furlong from the former, and is** equipped with **a white light, which** emits flashes of two seconds' duration **at** intervals of five seconds. It is also provided with a fog-siren.

This is **a** dangerous **part of the** coast, where wrecks have **been** frequent. **The** *Varne* **shoal, since** 1860, has been indicated by **a lightship, which has a red** light, revolving every twenty seconds.

The *South Foreland* is the nearest point to the French coast, and on a clear night commands a distinct view of the lights **of Dunkirk, Calais, and** Boulogne. It is a noble promontory **of chalk**, with an elevation of

---

* On board this vessel, in **1875,** was displayed the **first** group-flashing floating light, showing three successive flashes, as above described.

three hundred feet above the sea; and is surmounted by
two lighthouses, of which the upper, erected in 1843,
on the site of a structure built by Sir John Meldrum
in 1634, is a square tower of stone sixty feet high, and
the lower, three hundred and eighty-five yards distant,
an octagonal structure of stone, forty-nine feet high,
erected in 1846. Both were designed by the late James

THE LOWER SOUTH FORELAND LIGHTHOUSE.

Walker, and cost £4,409, 4s. 3d. Several important
experiments in the illumination of lighthouses have
been tried here. The strongest artificial light known
up to that time, and obtained from Professor Holmes's
magneto-electric apparatus, was introduced on December 8th, 1858. In 1861 were tested, but unsuccessfully,
the illuminating properties of the Drummond or lime-
light. In 1872, the electric system was permanently
established, Holmes's machines being at first adopted.

But the simultaneous discovery by the late Dr. Siemens and Sir Charles Wheatstone, **that** induced **electricity** could actively **be** generated without the use of permanent magnets, **led** to the construction of dynamo-electric machines; and these being **tested** at the **South Foreland**, in 1876–7, **it was conclusively** proved that they furnished a **much** superior **illumination** to that of any magneto-electric machine. Lastly, in 1884–5, exhaustive experiments were conducted here with a view to decide **whether gas or** electricity **is, for** lighthouse **purposes,** the better illuminant.

Experiments have **also** been made **at the South Foreland on the value of the siren and** the utility of **gun-discharges as signals in foggy weather.**

Few spots **along our island shores** are more dreaded by seamen, **or** invested with gloomier associations, than the Goodwin Sands, **which lie** outside the sheltered channel **of the** Downs, **at a distance from the shore of three to seven miles. They measure ten miles in** length **and nearly two miles in breadth, and at low** water present a surface so **hard and** firm that pleasure parties often land, and cricket matches have been played upon them. Anciently, they were known as Lomea (that **is, loam-ey).** Their modern name, according to an old tradition, originated in the fact that they represent a tract of land which **once** belonged to Earl Godwin, and **by** William the Conqueror was afterwards given to the monks of St. Augustin, Canterbury; but as they neglected to keep in repair the sea-wall, the whole tract was submerged in an irruption of the sea that occurred about 1099. It is, or was, a belief among seamen that a ship of the largest size, if she struck upon the Goodwins, would be swallowed up by the quicksands in a few days. There are, however, only fifteen feet of sand lying here upon a bed of blue clay.

Many a goodly vessel has perished on these fatal sandbanks. The worst calamity connected with them is the loss of a squadron of thirteen men-of-war, under Rear-Admiral Beaumont, in the night of November 26th, 1703. On December 21st, 1805, the *Aurora* transport was wrecked here, and three hundred soldiers and seamen perished. All on board the *British Queen* packet were drowned, on the 17th of December 1814. To prevent these disasters, several attempts were made to erect a permanent lighthouse; but in the earlier years of the present reign two beacons and a lighthouse were destroyed in succession. Their dangers are now indicated by four lightships,—the *East Goodwin*, the *South Sand Head*, the *Gull Stream*, and the *North Sand Head.*

The *East Goodwin* is moored in lat. 51° 18′ 30″. It has one green light, revolving, with fifteen seconds' interval, which is visible ten miles. This lightship was stationed here in 1874.

The *South Sand Head*, in lat. 51° 9′ 35″, was moored in 1832, and replaced in 1884. It has a white flashing light, visible ten miles, and carries also a fog-siren.

The *Gull Stream*, in lat. 51° 16′, fixed in 1800, has a

white light, revolving every twenty seconds, and visible ten miles.

The *North Sand Head*, in lat. 51° 19′ 23″, moored in 1703, has a white light, with three flashes every minute, visible ten miles.

LIGHTSHIP ON THE GOODWIN SANDS.

On board the *Gull Stream* lightship, some years ago, Mr. R. M. Ballantyne, well known as a literary caterer for the young, spent a night, of which he furnished an account to the Edinburgh *Scotsman*. A little before midnight, when he was lying half-asleep, and more than

half sea-sick, he was roused into activity and convales-
cence by a cry from the watch on deck to the mate,
"South Sand Head light is firing, sir, and sending up
rockets." The mate was soon on deck, followed by his
visitor. They found the two men on duty actively at
work ; the one loading the lee gun, the other adjusting
a rocket to its stick. The mate's few rapid questions
elicited all that was needful to be known. The flash
of a gun and the glint of a rocket from the South Sand
Head lightship, about six miles distant, had indicated
that a vessel had got upon the fatal Goodwins. "While
the men spoke," says Mr. Ballantyne, "I saw the bright
flash of another gun, but heard no report, owing to the
gale carrying the sound to leeward. A rocket followed,
and at the same moment we observed the light of the
vessel in distress, just on the southern tail of the sands.
By this time our gun was charged and our rocket in
position. 'Look alive, Jack ; get the poker,' cried the
mate, as he primed the gun. Jack dived down the
companion-hatch, and in another moment returned
with a red-hot poker, which the mate had thrust into
the cabin fire at the first alarm. Jack applied it in
quick succession to the gun and the rocket. A blinding
flash and deafening crash were followed by the whiz of
the rocket, as it sprang with a magnificent curve far
away into the surrounding darkness. This was our
answer to the South Sand Head light, which, having fired three guns and three rockets to attract our
attention, now ceased firing. It was also our note

of warning to the look-out on the pier of Ramsgate Harbour. 'That's a beauty,' said our mate, referring to the rocket; 'get up another, Jack; sponge her well out, Jacobs; we'll give 'em another shot in a few minutes.' Loud and clear were both our signals, but four and a half miles of distance and a fresh gale neutralized their influence. The look-out did not see them. In less than five minutes the gun and rocket were fired again. Still no answering signal came from Ramsgate. 'Load the weather-gun,' said the mate. Jacobs obeyed; and I sought shelter under the lee of the weather bulwarks, for the wind appeared to be composed of penknives and needles. Our third gun thundered forth, and shook the lightship from stem to stern; but the rocket struck the rigging and made a low wavering flight. Another was therefore sent up; but it had scarcely cut its bright line across the sky when we observed the answering signal—a rocket from Ramsgate Pier."

Their work was done, and the mate went below, while the watch resumed their active perambulation of the deck. Mr. Ballantyne felt somewhat disappointed at the sudden termination of so exciting a scene. He was informed that the Ramsgate lifeboat could not well come to the rescue in less than an hour; and it seemed to him a terrible thing that human life should be held so long in jeopardy, and he asked himself if it were not possible that the delay might be prevented. There was no remedy, however, but patience; and he

wisely turned into his berth, with orders that he should
be called as soon as the lights of the steam-tug became
visible. It seemed but a few minutes after when the
voice of the watch was again heard, shouting, "Lifeboat close alongside, sir. Didn't see it till this moment.
She carries no lights." Mr. Ballantyne did not wait
for coat, hat, or shoes, but hastily scrambled on deck
just in time to see the Broadstairs lifeboat sweep past
before the gale. "What are you firing for?" cried her
coxswain. "Ship on the sands, bearing south," shouted
Jack, at the full pitch of his stentorian voice. The
boat did not pause. With a rush like that of a race-
horse it drove into the darkness. The reply had been
heard, and straight as an arrow the lifeboat made for
her quarry. Then, once more, silence reigned on board
the lightship. But not for long. "Tug's in sight, sir,"
exclaimed the watch; and soon the sturdy little steamer
appeared, with the Ramsgate lifeboat in tow far astern.
As she passed, the brief question and answer were
repeated; and Mr. Ballantyne had just time to observe
that every man, except the coxswain, lay flat on the
thwarts. And no wonder! It is no easy matter to sit
up in a gale of wind, with freezing spray and sometimes great billows sweeping over you. "They were
doubtless wide awake and listening; but, as far as
vision went, that boat was manned by ten oilskin coats
and sou'-westers. A few seconds took them out of
sight; and thus, as far as the *Gull* lightship was concerned, the drama ended." But next morning the

wrecked ship lay, bottom up, high on the Goodwin
Sands. It was the *Germania* of Bremen.

The lights of Ramsgate Harbour now rise into view,
and shortly afterwards we double the *North Foreland*, the extreme north-eastern point of Kent—the
Cantium Promontorium of Ptolemy. The lighthouse

NORTH FORELAND LIGHTHOUSE.

here occupies the site of a wooden structure, surmounted
by a glass lantern, built by Sir John Meldrum in 1634,
and burned down in 1683. An octagonal tower, two
stories high, was then erected, and on the top was set
an open iron grate or chauffer, fed with coals, the fire
being kept alive at night by the action of a pair of

bellows. In 1790 two stories of brick were added, and oil lamps introduced. In 1884 the building was restored and improved. It is now eighty-five feet in height, and provided with a white light, occulting every half minute for five seconds; is octagonal in shape, and built of stone. It was purchased in 1832, with the two South Foreland lighthouses, for £8,399, 16s.

Reference may be made to the *Nore* lightship (lat. 51° 29′) in the mouth of the Thames, because she is so well known—steamboat excursionists to Margate and Ramsgate, and passengers on board deep-sea steamers, sailing within sight of her where she lies at her lonely moorings. She carries a white light, revolving every thirty seconds, and visible for ten miles. Stationed at the Nore in 1732, she was the first of these aids to navigation ever used. Her illuminating apparatus consisted of a small lantern provided with flat wick oil lamps; but afterwards Argand burners, aided by paraboloidal silvered reflectors, were introduced, each reflector and lamp being properly gimballed to insure the horizontal direction of the beam during the pitching and rolling of the vessel.

The intricate navigation of the estuary of the Thames is also assisted by the *Tongue* lightship, on the north side of the Tongue sandbank, which displays a red and white flash in quick succession every half minute, and is provided with a fog-siren; the *Princes Channel* lightship, which has a red light, revolving every twenty seconds; and the *Girdler* lightship, lat. 51° 29′, with a white light, revolving at half-minute intervals. In Sea Reach we find pile lighthouses erected on the *Chapman* bank and *Mucking* flat—the former, seventy-four feet high, rebuilt in 1881, showing an occulting light, with white and red sectors; and the latter, seventy-one feet high, also rebuilt in 1881, displaying the same light. Iron tressle-work towers, painted red, were erected in 1885 at the end of *Broadness* and *Stoneness*. Each is forty-four feet from base to vane, and thirty-eight feet above high-water mark to centre of lantern. The *Mouse* lightship, moored in 1838, lat. 51° 32′, is at the west end of the sandbank so called, and hoists a green light, revolving every twenty seconds.

On the south-east edge of Maplin Sands was erected in 1838-41, from the designs of Mr. James Walker, C.E., the *Maplin Lighthouse*, at a cost of £4,948, 6s. 9d. It stands upon nine piles of wrought iron, each twenty-six feet long and five inches in diameter; these are screwed into the sand fourteen feet six inches deep, and secured (on Alexander Mitchell's system) by cast-iron screw-blades, each four feet in diameter. One pile forms the centre of an octagon, the others being placed one at each of the eight angles. To the tops of the piles are firmly fitted hollow columns of iron, the central being perpendicular and the others bent, so that they incline inwards; all are braced together by radiating, diagonal, and horizontal rods. Each terminates at the top in a socket, into which is fitted a timber-post, about one foot square. The posts are braced together like the columns, and support the platform on which is constructed the lighthouse with its lantern.

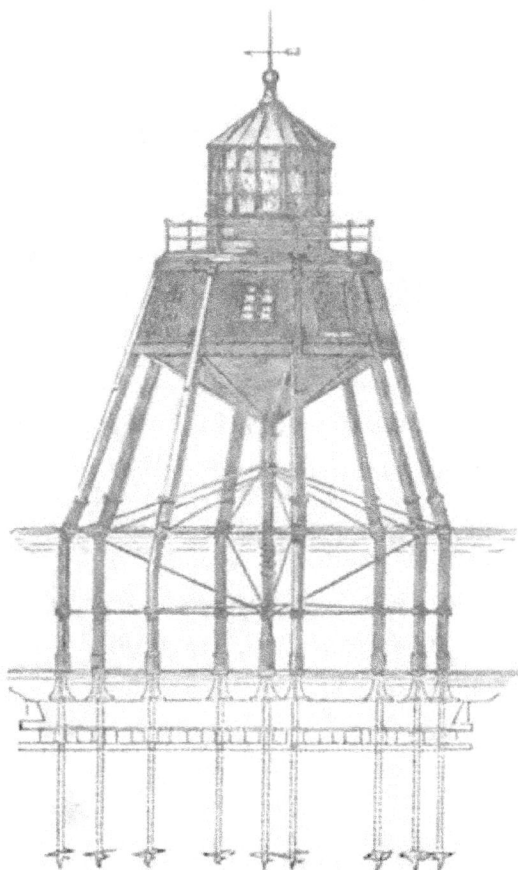

MAPLIN SANDS LIGHTHOUSE.

The principal measurements are: depth of screw-blades below the sand, $14\frac{1}{2}$ feet; depth of the screw-blades below spring-tide low-water mark, 21 feet; rise of spring-tides, 15 feet; height from spring-tide high-water mark to floor of house, 20½ feet; height from high-water mark to floor of light-room, 29½ feet; height from high-water mark to lamp, 38½ feet; height from high-water mark to top of vane spindle, 54 feet; diameter of floor of house, 27 feet; diameter of platform, 21 feet; diameter of light-room, 12 feet.

A lighthouse of this construction is of special utility in localities where the shifting and uncertain bottom does not permit of the erection of a mass of masonry, and the light is not required to extend to any great distance. The piles offer but slight resistance to the waves, which pass through the open spaces, and rise no higher than they do at sea.

The Maplin shows an occulting red light, with white sector, at half-minute intervals, visible for ten miles.

The *Swin* is marked by a lightship with a white revolving light. On the *Gunfleet* sand is a screw-pile lighthouse, dating from 1856, which is seventy-two feet high, built of iron, and six-sided; was designed by James Walker, and cost £14,567, 7s. 5d. It has a red light, revolving every half-minute. The *Long Sand* lightship, in lat. 52° 48′ 16″ (stationed in 1883), carries a white light, double-flashing—a flash of three seconds, eclipsed for six seconds; a flash of three seconds, eclipsed for thirty seconds. On the east side of the *Kentish Knock* sandbank, the lightship, first moored there in 1840, exhibits a white light, revolving every minute. Between the Gunfleet pile lighthouse and the

Long Sand lightship is placed, in lat. 51° 30′ 40″, the *Sunk* lightship, showing a red and white light alternately, revolving in forty-five seconds. To this last-named vessel an electric cable has for some time been attached, connecting it with the post-office at Walton-on-the-Naze, and in February 1887 the Board of Trade appointed a committee (the Earl of Crawford; Colonel King Harman, M.P.; Sir Edward Birkbeck, M.P.; Rear-Admiral Sir George Nares; Mr. Cecil Trevor, assistant-secretary to the Board of Trade; Mr. Thomas Sutherland, M.P.; Captain J. Sydney Webb, deputy-master of Trinity House; Captain Hozier; and Mr. Alfred Holt) to inquire whether a system of electrical communication should be applied to outlying lighthouses, whether the experience gained by laying the present cable to the Sunk light-vessel has proved sufficient to justify the cost, whether the system should be further extended, and whether it is desirable that any of the stations to which it is applied should also be used as signal-stations for commercial purposes. It might be thought that the electric connection of our lighthouses and lightships with the nearest lifeboat station would prove of greater service in the rescue of life and the relief of suffering than the most elaborate apparatus of gun and rocket signals, as it would virtually be independent of all conditions of the atmosphere; but the decision of the committee was not in favour, however, of any extended application of the system, owing to certain practical difficulties in the way of its working.

The foregoing enumeration will impress the reader with an idea of the fulness and efficiency with which the navigation of the Thames estuary and its approaches is protected. We must add to it due record of the *Galloper* light-vessel, in lat. 51° 45′, which has a fixed white light, set horizontally; the lights at Clacton-on-Sea, Harwich, and Woodbridge Haven; the *Cork* light-ship, in lat. 51° 55′ 55″, displaying a white light, revolving every half-minute; the *Shipwash* lightship (1883), with its triple-flashing half-minute light; and the lighthouses on the low flat projection of Orfordness. The higher of these is a circular tower, ninety-nine feet high, erected in 1792; the lower, a quaint sixteen-sided building, seventy-two feet high. They are distant from each other about three-quarters of a mile (1,439 yards).

Passing Southwold and Pakefield—*Lowestoft*, where a lighthouse was erected on the cliff as early as 1609 (rebuilt 1628 and 1676), and one on the low ness or projection in 1609, restored in 1832, and improved in 1881; the former with a revolving white and a fixed red light, the latter displaying an occulting light, with red and white sectors—we come to three lightships in succession: the Corton, outside Corton Sands, with a red light, revolving every twenty seconds; the Hewett Channel or St. Nicholas Gat, in lat. 52° 34′ 40″, exhibiting a fixed white light, and, near the vessel's stem, a red light, flashing every ten seconds; and the Middle Cross Sand, lat. 52° 38′ 15″, with its double-flashing white light. A singular accident occurred on board the St. Nicholas lightship on Saturday morning, February 19th, 1887. The attention of the crew having been directed to a vessel in dangerous proximity to Scroby Sand, a gun was fired to warn those in charge of her. As it did not seem to have the desired effect, a rocket was sent up. Some sparks descending from it unfortunately fell into a store of gunpowder on the deck of the lightship, and an explosion occurred, which, however, serious as it was, caused no loss of life. Two or three of the crew were slightly injured, and one received wounds which necessitated his removal to the hospital.

Leaving behind us the lights of Yarmouth, we sight the *Cockle* lightship, moored on the eastern side of the north entrance to the Yarmouth Roads—a frequent rendezvous of our fleet in the great Revolutionary War. The white light revolves every minute, and is visible for ten miles. When she was visited by the Royal Commissioners in 1859, the master stated that she had twice drifted from her moorings in a north-westerly gale—namely, in 1849 and 1856—but on each occasion was brought up almost immediately with a spare anchor, or she would have been lost on the sands. In similar weather, with a lee-tide, the master and one of the crew always remained on deck, with axes ready, to cut the spare anchor adrift.

"This light is revolving. She has four reflectors, which were in such good order that the lamplighter was requested to show his **process of cleaning them.** He first put on a canvas apron; he then selected **from** a particular **box a** clean **white** cloth, with **which he wiped the inside** of the reflector, which he **held against** his breast, carefully avoiding **to touch the silver with** his hand. **He next** dusted **some** rouge over **the** silver from **a linen bag,** which acted **as a kind** of sieve, and **the cleaning was** finished with **a** leather, taken from another **box used for that** purpose only. **There was** nothing **peculiar in the process,** and the **man** could **see no** reason **why reflectors should** be more scratched **at sea than** on land. **Many** vessels run foul **of the** *Cockle.*"

At *Winterton* (lat. **52° 43′), the** lighthouse, **erected in** 1790, is sixty-nine feet **in height, and built of** brick, **coated with** cement.* **It shows a fixed white** light, visible at **fourteen miles. We** particularize this spot simply because **the building (erected** about 1615) **which preceded the present is mentioned in Defoe's immortal** "Robinson Crusoe." **The young** sailor is describing the escape **of his** comrades **and himself from their** foundering **ship.** "Partly **rowing and partly driving,** our boat went away **to the northward, sloping towards** the shore almost as far as Winterton **Ness......While we** were in this condition, the **men** yet labouring at the **oars to bring the boat** near the **shore, we could see, when, our boat mounting the waves,** we were **able to**

---

* It was purchased (with **the two Orford lights) from** the lessee in January 1857, for £37,806.

see the shore, a great many people running along the
shore to assist us when we should come near. But we
made but slow way towards the shore ; nor were we
able to reach the shore, till, being past the lighthouse
at Winterton, the shore falls off to the westward
towards Cromer, and so the land broke off a little the
violence of the wind."

As we continue our course northward, we descry
quite "a lane" of lightships between Winterton Ness
and the mouldering cliffs of Cromer, which encircle the
dangerous bay expressively named "The Devil's Throat."
We pass, in succession, the *Newarp*, near North Cross
Sand, with a triple-flashing white light (every minute);
Smith's Knoll, one mile eastward of the shoal, with a
red and white double-flashing light, and a fog-siren ;
the *Would*, off the south end of Hasborough Sand,
with a white light, flashing every five seconds; the
Hasborough, near the north end of the sandbank, with
a fixed white light, set horizontally ; and the *Leman
and Ower*, between the two sandbanks so called, showing two flashes of white light every half minute. The
Hasborough lighthouse must be mentioned here : it
stands on the cliff, in lat. 52° 40′ 12″; was erected
originally by Wilkins and Norris, in 1791, and rebuilt
by Sir James Douglass in 1884; is eighty-five feet
high, and displays a white light, occulting every half
minute.

The *Cromer* lighthouse is an octagonal structure of

brick and stone, fifty-eight feet in height, erected by Mr. James Walker in 1823, and furnished with a revolving white light, visible twenty-three miles. The cliffs are here about two hundred feet above the sea. A coal-fire was lighted at this station in 1729, and replaced by oil lamps in 1792.

On *Hunstanton Point* a coal-fire was established in 1665. The present building, designed by James Walker, is a circular tower of brick, coloured white, and sixty-three feet from base to vane, with a white occulting light, visible for twenty-four seconds, then eclipsed for two, visible for two, and eclipsed for two. Visible sixteen miles. Cost of tower and residence, £2,696, 13s. 4d.

The *Lynn Well* lightship, off the Hook of Long sandbank, has a white light, with a flash every ten seconds. The *Bar Flat*, first moored in 1878, was built for the King's Lynn corporation, at a cost of upwards of £2,000. The hull is of iron, and divided into three water-tight compartments. Upon her one mast, at a height of forty feet above water-level, is the lighting apparatus, consisting of dioptric lanterns arranged on a triangular frame, the three lights apparently blending into one at a distance, and being visible at all points of the compass within a radius of seven nautical miles.

Crossing the Wash, we pass the mouth of the Witham, and descry the twinkling lights of the busy port of Boston. Beyond Skeyness may be seen the flashing light, white and red alternately, of the *Dudgeon* lightship (lat. 53° 13′ 40″). Then the *Inner Dowsing* and the *Outer Dowsing* lightships, the former with a green light, revolving every twenty seconds, and the latter with a red light, revolving every half minute, and a fog - siren, uttering its gruesome blasts every two minutes.

We now arrive off the estuary of the Humber, one of the great water-ways of the kingdom, possessing a navigable channel for large ships for twenty miles up to Hull. We may take as its northern boundary the cliffs of Spurn Head, in lat. 53° 34′, and as its southern, Faxfleet Ness, beyond Whitton. Two lighthouses are situated on *Spurn Head*, erected in 1776 and 1852 respectively. The high tower, designed by John Smeaton, and built of red brick, is enclosed within a circular wall, along with the keepers' residence, a paved court, and neatly-kept gardens. It measures one hundred and twelve feet from base to vane. Lights were first shown here in April 1675. The Trinity House bought up the buildings and the tolls in 1841, for the large sum of £309,531, 4s. The low tower, one hundred and fifty-eight yards distant, designed by James Walker, and built of brick, is seventy-six feet high, and coloured white. It stands within high-water mark, on a foundation of piles and concrete. Both are situated on a long spit of sand overgrown with bent. To seaward, a strong chalk wall has been raised to prevent the encroachment of the waves. The dwelling-houses are large, roomy, and kept in excellent order, the light-towers scrupulously clean, every bit of metal shining like a schoolboy's newly-washed face. There are three keepers, and the one on watch walks from tower to tower during the night. The high light is red and white, occulting every thirty seconds; the low, a fixed white.

The *Spurn* lightship, off the point, carries a revolving white light; the *Bull Sand*, a fixed white light; the *Grimsby*, a fixed red. There are two lights at *Narsham Booth*, erected in 1880. At *Killingholm*, the high lighthouse, painted red, is seventy-seven feet high; north tower, also red, forty-five feet; and south-east tower, white, forty-five feet high. These are furnished with white fixed lights, visible eleven miles, which lead over the dangerous Holme and Paull sands. The old Paull lighthouse is now disused; but at *Thorngumbald Clough*, half a mile to the south, two lighthouses were erected in 1870—the high lighthouse, red, fifty feet high, the low lighthouse, a circular tower, yellow, thirty feet high. These show a fixed white light. There are also lights at Salt End, Storsle, Chaldersness, Ferrily Shine Haven, Winteringham, and Brough. There is a lightship (white and red lights) off Whitton. The lights at Walker Dyke and Flaxfleet do not call for notice.

Keeping along the Yorkshire coast, with its gradually rising line of cliffs, we pass Withernsea, Hornsea, and Bridlington, thirty miles from Spurn Point, before we arrive at the bold and rugged promontory of *Flamborough Head*, with its precipitous cliffs and dark caves. Its summit is crowned by a circular stone tower,* eighty-seven feet high, with the centre of its lantern two hundred and fourteen feet above high-water mark. Its illuminating apparatus exhibits two white faces and one red, each half a minute. During thick weather an explosive rocket, reaching an altitude of about six hundred feet, is discharged every ten minutes. Flamborough Head is in north lat. 54° 7′, and west long. 0° 5′ 0″.

Whitby, besides its harbour lights, shows a couple of powerful fixed white lights, visible for twenty-three miles, on the summit of the Ling Hill, in lat. 54° 28′ 40″. The two are situated north and south, at a distance of two hundred and fifty-eight yards, and were first shown on 1st October 1858. The buildings, designed by Mr. James Walker, are octagonal in shape and built of stone, the one sixty-seven feet and the other forty-six feet high, while the elevation of the centre of the lantern above high-water mark is, in each case, two hundred and forty feet. Cost £5,256.

On the north side of the mouth of the Tees, lat. 54° 40′, stand two circular stone lighthouses, the high light half a mile inland, and the low light on the shore, north of *Seaton Carew* (eleven hundred and eighty-three yards from the other). Each is seventy feet

---

* Situated about four hundred yards from the extreme point of the Head. Designed by Mr. S. Wyatt, the architect. Cost, with adjoining buildings, £4,099.

high. The high light is white, the low, red; both are fixed, and were first exhibited in 1839.

On the Heugh at *Hartlepool*, a light-tower of solid freestone (coloured yellow) was erected in 1847, from the designs of Stephen Robinson. It is seventy-three feet seven inches in height, and exhibits a white and a red light, the latter twenty-four feet lower than the former. The white light is visible for fifteen miles, and the red light for four miles. Cost £3,644, and lantern (which is fourteen feet in diameter by ten feet in height), £460.

At *Seaham*, on Red Acre Point, a stone building was erected in 1831. It is furnished with an upper light, fixed white, and a lower one, red, revolving every half minute. It was built by William Chapman, C.E.

Beyond Sunderland, and the mouth of the Wear, we come to *Souter Point*, in lat. 54° 58′ 10″, and long. 1° 21′ 30″, where a handsome new lighthouse of the first class, seventy-six feet high, was "inaugurated" in 1871, and equipped with the electric light, Holmes's "alternating current" machines being made use of. This was the third station at which electricity was made available for lighthouse illumination. Two lights are displayed in the lantern—a white light, with flashes every half minute; and from a window, twenty feet lower, a fixed white light, with red sectors.

Coquet Island, in lat. 55° 20′ 6″, on the coast of

Northumberland, is a tract of rich green pasture, once the site of a fortress, now, on its south-western shore, of a lighthouse of substantial construction—a square white tower of sandstone, with turreted parapet, and walls three feet nine inches thick, seventy-two feet in height, and exhibits white and red lights, intermittent, suddenly disappearing for five seconds every minute. Designed by Mr. James Walker, and first lighted 1st October 1841. Cost (with keepers' dwellings) £3,268, 5s. 2d.

Passing Warkworth, famous in ballad history, we sight the lofty basaltic headland of Bamborough Castle, off which lie the Farne and Staples islets, seventeen in number, forming two picturesque groups, inner and outer, whither the sea-birds resort to breed and bring up their young. These islets render navigation, however, both difficult and dangerous, and as an assistance to the mariner a lighthouse was erected on the south-west point of *Farne Island,** the largest of the cluster, as early as 1776, when a coal-fire blazed on its summit; rebuilt by Mr. Joseph Nelson in 1809–10, as a structure forty-three feet high, which now displays a flashing light—two white flashes and one red, each lasting twenty seconds, followed by an eclipse of the same duration, visible fifteen miles. In 1811, a lighthouse, designed by Mr. Joseph Nelson, was erected on the

* A patent was granted as early as December 1st, 1673, to Sir John Clayton and George Blake to erect lighthouses on the Farne Islands. The Trinity House purchased the two Farne lighthouses and the Longstone, in December 1824, for £36,445, 13s. 2d.

north-west side—a low, white octagonal building of rough stone, only twenty-seven feet high, with a fixed white light, visible twelve miles. In 1826 it was found advisable to construct another on the *Longstone* rock, which lies just outside the Farne group, about six miles from the mainland, and, as it is only four feet above high-water mark, is swept in every gale with fierce rushing drifts of spray and foam. The tower, coloured red for the sake of distinction, is eighty-five feet in height from base to vane. It carries a white light, revolving every thirty seconds, and has of late been furnished with a fog-siren, which, in bad weather, utters two blasts in quick succession every two minutes. Architect, Joseph Nelson. Light first exhibited 15th February 1826.

In 1838 the keeper in charge was named Darling; and he had a daughter, Grace, who lived with her parents on this lonely, water-bound islet—a quiet, unassuming, well-principled girl, who, to all appearance, had nothing of the heroic in her. It happened that, on the 6th of September, the steamer *Forfarshire*, from Hull to Dundee, was driven by a violent storm on the rocks in the neighbourhood of the Longstone, and soon became a total wreck. At the time most of the passengers were below, and as the vessel filled rapidly, they were, unhappily, drowned in their berths. A portion of the crew contrived to escape in a boat; the rest, with the surviving passengers, sought a precarious asylum in the fore part of the shattered ship, over which the swirling waters poured with almost irresistible violence. Desperate as was their situation, they clung to life with all the energy they could command. Their hearts sank within them as they felt their strength ebbing away, and their cries seemed to be silenced by the roar of the billows; but, happily, they fell upon the quick ears of the daughter of the Longstone lighthouse-keeper. Without hesitation she roused her father, and as soon as the first faint gleam of morning broke on the eastern horizon, launched the lighthouse coble, to proceed, if possible, to the rescue of the sufferers. The tide ran with such violence, and the wind blew so tempestuously, that an attempt to reach the wreck seemed impracticable as well as dangerous; and the old man at first recoiled from rushing, as he considered it, upon certain death.

After watching the wreck for some time they made out that living beings were still hanging to it; and the heroic girl, with a courage that must be called chivalrous, suddenly seized an oar and leaped into the coble. This was sufficient; her father followed, and the two were soon pulling vigorously towards the rock.

By a desperate exertion, the lighthouse-keeper contrived to board the shattered vessel, while his daughter skilfully backed the coble to prevent it from being dashed into fragments. By watching for occasional lulls in the fury of the gale, they at length succeeded in taking off nine persons, and conveying them safely to the lighthouse, where Grace Darling patiently and gently waited upon them for three days and three nights. A four-oared boat from North Sunderland reached the wreck later in the day, but found no living person on board it. In attempting to return, her crew were compelled to put in at the Longstone lighthouse, and to remain there a couple of days and nights in a temporary shelter, the waves occasionally bursting in and driving them into the lighthouse-tower, which was already occupied by Mr. and Miss Darling, and the seamen and passengers they had so gallantly saved.

Mr. Walter White, in his lively book of travel, "Northumberland and the Border," describes a visit which he paid to the lighthouse some time after Grace Darling's premature death. Her sister, he says, a quiet-looking, middle-aged woman, of respectful manner, welcomed him and his friends, and led the way up to the sitting-room. It had a comfortable look and something more, with its collection of books, natural curiosities, engravings of Grace's memorable exploit, and family portraits. Presently old Mr. Darling, Grace's father, made his appearance, and proudly exhibited a copy of the letter which he wrote the day after the fatal wreck, to inform the secretary of the Trinity House of the adventure which has made his daughter's name illustrious among those of heroic women.

Mr. White afterwards went up to the lantern, and out upon the gallery, whence, as the tower is sixty-three feet high from the rock, an ample and richly-

varied view over the islands is obtained. The old
man pointed out to him the course which he and his
daughter took on their way to the wreck; and explained that his wife had helped to launch the boat,
that Grace knew how to handle an oar, but that to
pull half a mile or more through a furious sea was
no easy task for a girl. Nor did he know how they
should have got the boat back to the lighthouse against
the tide, had not some of the men whom they saved
been able to row.

"My eye," says Mr. White, "roved over the scene as
we talked, looking down on the twenty-seven isles and
islets of the Farne group as on a panorama. There,
nearly six miles from the shore, the isolation appears
somewhat awful; and we may think that the courage
of the residents was tried in the storm some years ago
which brought in such tremendous waves that they
had to seek the upper chambers of the lighthouse. The
Longstone, rising but four feet above high-water mark,
is swept by every gale with fierce drifts of spray and
foam; hence its vegetation is of the scantiest, including
but five kinds of plants, among which the sea mat-
grass predominates. Far different from the present
scene: for now children are at play on the rock; the
poultry look as if the weather were always fine to
them; and two boys, who have just come in from
fishing, are cleaning and washing their capture, throwing the refuse over the stern of their boat, and the
young gulls, hovering round, dart down with a shriek

and seize the dainty morsels sometimes before they touch the water."

Grace Darling lived but a few years after the event which made her famous. She lies buried in Bamborough churchyard, where a monumental tomb was erected to her memory in 1846. On this her effigy is seen in a recumbent position, her hands crossed, and an oar by her side which the right arm embraces. The inscription was written by Wordsworth :—

> " The maiden gentle, yet at Duty's call
> Firm and unflinching as the lighthouse reared
> On the island-rock, her lonely dwelling-place ;
> Or like the invincible rock itself that braves,
> Age after age, the hostile elements,
> As when it guarded holy Cuthbert's cell."

When the Royal Commissioners visited the Longstone in 1859, they found William Darling still acting as keeper, with his son as assistant. The whole establishment was in excellent order, and filled with memorials of Grace. The old man repeated to his visitors the story of the wreck, and from the tower pointed out the different localities. He told them that his daughter had died of a decline, and that in his opinion her end had been accelerated by "anxiety of mind"—in other words, " so many ladies and gentlemen came to see her that she got no rest."

In July 1843, the *Pegasus* steamer, plying between Leith and Hull, was wrecked in this dangerous neighbourhood. She left Leith at half-past six in the even-

ing of the 19th, and at half-past ten was sighted off
Berwick Harbour. At midnight she passed Holy Island,
and steered for what is called the Fairway, a channel
between the Farne group of islets and the mainland.
It is about a mile wide, and deep enough for the largest
ships; but its navigation is rendered difficult by the
numerous sunken rocks that lie in all directions.
Lights and buoys, however, are stationed at every
critical point; and it appears that on the night we
are speaking of the buoy which marked the Goldstone
rock, as well as the Farne lights, were clearly discernible. Not the less, the *Pegasus*, while steaming
at full speed, struck upon the Goldstone, which at the
time was under water. Her captain and officers seem
then to have lost their heads; the greatest confusion
prevailed on the wreck; and of all on board only six
were saved.

SCOTLAND.

Passing Berwick, we enter upon a survey of the east
coast of Scotland at *St. Abb's Head*, a huge isolated
mass of trap rock, with a precipitous seaward front
nearly two hundred and fifty feet high. The lighthouse erected here in 1862 is twenty-nine feet from
base to vane, exhibits a flashing white light every ten
seconds (visible twenty-one miles), and is furnished
with a fog-siren.

We sail past Coldingham; Dunbar, with the ruins
of its historic castle, and its memories of the great
battle in which Cromwell defeated the Scotch army;

North Berwick, easily recognized by the pyramidal height of "the Law" which overlooks it; and on the rocky isle of *Fiddra* descry the tall brick tower, fifty-six feet high, erected in 1885 to display a white light, with two flashes every fifteen seconds—the first group-flashing light in Scotland—visible in fair weather over a range of seventeen miles. The Firth of Forth is admirably lighted throughout the whole extent of its tidal way. On the left bank we trace the lights of Port Seton, Cockenzie, Fisherrow, Leith, Newhaven, Granton, the island of Inchkeith, the Oxcar Rocks, Newhalls, South Queensferry, Bo'ness, and Grangemouth. Returning along the Fifeshire shore, we pass Alloa, Charleston, North Queensferry, Inchgarvie (where the famous Forth Bridge, the greatest triumph of engineering skill which the world has yet witnessed, now crosses the noble river), Inverkeithing, St. David's, Burntisland, Kirkcaldy, Dysart, West Wemyss, Buckhaven, Elie, St. Monans, Pittenweem, Anstruther, and Cellardyke. With the exception of that upon Inchkeith and on the Oxcar Rocks, to which we shall hereafter refer, these are harbour lights, which do not fall within the scope of the present volume.

We continue our imaginary voyage across the broad estuary of the Forth, pausing at the long and narrow rocky mass of—

The ISLE OF MAY, which commands the channel to the well-known secure anchorage at St. Margaret's Hope. So early as the year 1621 we find the better navigation of the Forth made a matter of high consideration; and we find that the coal-masters on both sides of the river, "for the credit of the country and the safety of strangers trading to them for coal and salt, undertook to put marks and beacons on all the craiges and blind rocks within the Forth, above Leith Road, upon their own charges." Not, however, until fifteen years later was any attempt made to provide for the guidance of mariners at night; and then the Scottish Parliament granted permission (April 22, 1636) to James Maxwell of Innerwick, who was one of the king's bedchamber, and James Cunninghame of Barnes "to erect and maintain ane light upon the Isle of May," the toll from passing vessels to be "twa shillingis Scottis, to be payit upon ye tun of all shippes and veschelles coming within Dunnoter and St. Tobes Heid be natives, and four shillingis money be strangers for ilk veadge." The light (a coal-fire blazing in an open chauffer) thus kindled in 1636 continued to burn nightly till 1816. The tower and keepers' houses were probably built out of the ruins of the ancient monastery that had formerly flourished on the island. The tower, on the summit of which blazed the great bonfire nightly, was forty feet in height, the fuel being hoisted up by means of a pulley and box fitted outside it. The light consumed about two hundred tons, and latterly four hundred tons, annually, and was one of the best in the kingdom. But the toll exacted from shipping for maintaining it was thought too high by the Convention of Royal Burghs, and was accordingly reduced to "auchtein pennis Scottis for natives, and three schillingis Scottis for stranger" ships.

The lordship of the island and the valuable income from the toll on passing ships passed through various hands until it came to the Duke of Portland by his marriage to Miss Scott of Balcona. When the Northern Lighthouses Board was established in 1786, the shipping interests evinced a strong desire that the Isle of May light should be brought under the jurisdiction of that Board, and also that the penalty on English ships as "strangers" should be abolished. The Edinburgh Chamber of Commerce petitioned Miss Scott to improve the light, as limekilns and other fires on the Fifeshire and Berwickshire coasts were apt to be mistaken for the Isle of May light. The Trinity House of Leith, at a later day, also urged that oil should be substituted for coal; but the recommendation was disregarded, and the sole improvement vouchsafed was the enlargement of the chauffer to thrice its original dimensions. The light, however, still remained uncertain in its character, subject to the conditions of the weather, and liable to be mistaken for lights ashore. In violent gales the fire kindled only on the leeward side, and the keeper was accustomed to thrust his arm through the windward bars of the chauffer to steady himself when supplying fresh fuel, so that scarcely any light was visible in the direction in which it was most needed. Matters were at length brought to a climax

by the wreck of a couple of frigates in 1810, which
was mainly due to the inefficiency and the misleading
nature of the Isle of May light. Lord Melville, First
Lord of the Admiralty, then proposed its transfer to
the Commissioners, and negotiations were opened up
for the purchase of the light and of the Duke of Portland's interest in the toll on passing ships. In 1814
this purchase was authorized by an Act of Parliament,
which at the same time abolished the extra tax on
English shipping. For the sum of £60,000 the light
was transferred to the charge of the Northern Lighthouses Board, which proceeded to erect a substantial
tower, seventy-eight feet high, from which, on February
1, 1816, a new and powerful apparatus of Argand lamps
and silver-plated reflectors diffused its welcome radiance
across the waters. In 1844 a smaller tower was built
on the north-east side of the island, which directs a
broad beam of light towards North Carr Rock, to
guide vessels clear of that danger when doubling Fife-
ness. In 1887, however, a light-vessel was moored off
this rock, in lat. 56° 17′ 30″, with a white fixed light,
visible for ten miles.

The progress made in lighthouse illumination has
been very completely illustrated on the Isle of May.
First, there was the open coal-fire, with its continually
varying character and exceedingly limited efficiency.
Next, as we have seen, came reflectors and Argand
lamps, fed with sperm oil, and enclosed in a glazed
lantern. Fifty years ago these gave way to a dioptric

apparatus, with a central flame from a four-wick lamp. Colza oil was afterwards substituted for sperm, and, in 1874, paraffin for colza, with the result that, at about one-fifth of the cost, the intensity of the flame was increased fully ten per cent. And lastly, in 1886, the electric light was introduced. Now, the power of the beam of light given out by the paraffin was equal to nine thousand five hundred candles; whereas that of the electric light, as now shown, is equal to about three million candles, or more than three hundred times the intensity of the paraffin light, and, moreover, in foggy weather it can be increased, if desired, to six million candles. This beam of light is one of the most intense, if not *the* most intense, exhibited as yet from any lighthouse, all the latest improvements, not only in the electric but in the optical apparatus, having been adopted. The erection of suitable buildings, dioptric apparatus, steam-engines, and other appliances occupied a period of eighteen months. The buildings consist of dwelling-houses for the engineer and two keepers additional, an engine and boiler-houses, workshop, coal-store, and offices. The walls are constructed of bricks set in Portland cement and white-washed; the roofs, as well as those of the engine and boiler-houses, are of iron buckle plates covered with Portland cement and Val de Travers, supported on iron beams. The material employed for the other buildings is cement rubble concrete.

The new dwellings and engine-house are situated in a ravine or gully which traverses the island diagonally, at some distance from the light-tower. A small lake at the northern extremity of this gully has been converted into a reservoir, and the water is used for condensation purposes. In the engine-house are two engines, each of sixteen horse-power, for driving the electric machines, and also the two machines. As a rule, only one engine and one machine are used at a time—the spare engine and machine being provided to guard against accident, or to be brought into use along with the other in very dense, foggy weather. The machines are "alternate current" magneto-electric machines of the largest size hitherto constructed, having sixty permanent horse-shoe magnets arranged round a centre, in twelve sets of five each, all with their poles towards the centre. In the space left between the poles is a central shaft, on which are mounted five rings, on each of which are wound coils of wire, forming the armature, and this revolves within so as nearly to touch the poles of the permanent magnets. The armature is driven by a belt from the engine, through a counter-shaft, at a speed of ten revolutions per second; and each coil as it passes the poles of the permanent magnets produces an alternate positive and negative current of electricity. The pulsations of electrical agency are taken off by collectors at each end of the central shaft, and conveyed to the lighthouse-tower by copper rods one inch in diameter. These rods or conductors are placed in a groove in a low wall of cement rubble concrete, which connects the engine-house and the tower. Alongside this wall a pathway has been formed. The engine-house and tower are also placed in communication by telephone. The conductors are led up the tower by the lamps inside the dioptric apparatus.

The electric lamps are the best extant. One is always in its place inside the dioptric apparatus, while another is ready, with the carbons adjusted to the correct height, to be shunted into the focus of the apparatus when required. The carbons are fully one and a quarter inch in diameter, and have a small core of pure graphite running through the centre. These improved carbons never fail to burn with great regularity and permanence. At first it was intended to use continuous-current electrical machines, but a trustworthy arc lamp proved more difficult of construction than was imagined. The dioptric apparatus is novel of construction, on the condensing principle so successfully introduced into the Scottish lighthouses by Messrs. Stevenson thirty years ago. The Isle of May apparatus is so designed as to give a group of four flashes in quick succession, with intervals of darkness for half a minute, and condenses the light which would spread over forty-five degrees into three degrees or fifteen times. It has also been so arranged as to admit of the beams of light being dipped, to show near the lighthouse during dense fog, instead of being sent to the horizon.

We may here bring together a few facts in illustration of the gradual lighting of the Scottish coast. One of the earliest lighthouses erected seems to have been that of Grass Island, in Harris, first shown on October 10th, 1789, on which day a light was also exhibited at Pladda, a small island south-west of Arran, in the Firth of Clyde. As a guide to the passage of the Pentland Firth, a lighthouse was erected on the Pentland Skerries in 1794. A light was shown on Start Point, in the island of Sanday, on the 1st of January 1806; on the Bell or Inchcape Rock, February 1st, 1811; on the Skerryvore, February 1st, 1844. Of late years, under the active direction of the Messrs. Stevenson, a family with a traditional repute as lighthouse engineers, the shores of Scotland, like those of England, have completely been surrounded with efficient protection for the mariner.

From the Isle of May a short voyage up the pleasant Forth, passing the Fiddra light, brings us to the island of *Inchkeith*, which lies nearly opposite the town of Portobello on the south shore and Burntisland on the north. A white light, revolving at intervals of a minute, and visible one-and-twenty miles, has been shown here since 1804.* The tower, designed by Mr.

---

* Inchkeith was originally a *fixed* light, and when it was changed to a *revolving* one, a certain old lady, it is said, who had amused many a sleepless hour by watching it, was much puzzled by its successive illumination and eclipse; so that in the morning she expressed great compassion for the light-keeper,—"No sooner was his lamp lighted than it went out; and if it had been lighted once, it had been lighted a hundred times!"

Thomas Smith, is a handsome and well-built structure, forty-five feet in height, with its base one hundred and seventy-five feet above the sea. The dwelling-houses are very comfortable, and kept with a good deal of neatness. Farther up the river, in lat. 56° 1' 20", on the *Oxcar Rock,* a tower of stone and brick,

INCHKEITH LIGHTHOUSE.

seventy-three feet high, was erected in 1886. It shows a fixed red light, with white sectors, visible thirteen miles.

Returning to the mouth of the Forth, and directing our course northward, we come in sight of a lighthouse as famous in its way as the Eddystone—a lighthouse interesting from its tradition, situation, and difficulties of construction—the lighthouse on the BELL ROCK.*

The Bell Rock, formerly known as the Scape and the Inch Cape, is situated off the coast of Forfarshire, in lat. 56° 26′ north and long. 2° 23′ west, about twelve miles to the south-east of Arbroath. It lies in the direct track of vessels entering either the estuary of the Forth or that of the Tay, and therefore, in the old times, was much dreaded by the mariner as the most dangerous spot on the east coast of Scotland. In length it measures about one thousand four hundred and twenty-seven feet, of which at spring-tide ebbs a portion about four hundred feet long by thirty feet broad is uncovered to a height of about four feet; but at high water there is a depth of twelve feet over the whole of the reef. Its sides are thick with tangle and sea-weed; at low water it is frequently visited by seals, and, in larger numbers, by gulls, cormorants, and other ocean birds. The old name, "Inch Cape," probably refers to its situation as an "inch," or island, off the Red Head promontory; that of "the Bell Rock," by which it is generally known, either to its bell-like con-

---

* It was in the "Album" of this celebrated lighthouse that Sir Walter Scott, when he visited it in 1814, wrote the well-known lines:—

PHAROS *loquitur.*

"Far in the bosom of the deep
O'er these wild shelves my watch I keep,
A ruddy gem of changeful light,
Bound on the dusky brow of Night:
The seaman bids my lustre hail,
And scorns to strike his timorous sail."

figuration, or, more probably, to the circumstance that a bell with a float was fixed upon it by one of the abbots of Arbroath, in such a manner that it was tolled by the waves when they covered the rock. A tradition exists that the bell—"once upon a time"—was wantonly cut adrift by a Dutch pirate, whose vessel was soon afterwards dashed to pieces on this very reef; an incident which suggested to Southey his fine ballad of "Sir Ralph the Rover":—

> "No stir in the air, no swell on the sea,
> The ship was still as she could be;
> Her sails from heaven received no motion,
> Her keel was steady in the ocean.

> "With neither sign nor sound of shock,
> The waves flowed o'er the Inchcape Rock;
> So little they rose, so little they fell,
> They did not move the Inchcape Bell.

> "The pious Abbot of Aberbrothock*
> Had placed that bell on the Inchcape Rock;
> On the waves of the storm it floated and swung,
> And louder and louder its warning rung.

> "When the rock was hid by the tempest swell,
> The mariners heard the warning bell;
> And then they knew the perilous rock,
> And blessed the Abbot of Aberbrothock."

It is a bright summer's day, and Sir Ralph the Rover walks the deck of his ship with a heart "mirthful to excess." But his mirth was the mirth of a wicked heart, and for lack of other mischief he resolves to "plague the priest of Aberbrothock." He orders his

---

* The ancient name of Arbroath.

boat to be lowered, and rows for the Inch Cape Rock :—

> " Sir Ralph leant over from the boat,
> And cut the bell from off the float.

> " Down sank the bell with a gurgling sound,
> The bubbles rose and burst around ;
> Quoth he, 'Who next comes to the rock
> Won't bless the Abbot of Aberbrothock.' "

Then Sir Ralph sailed away and scoured the sea in quest of booty, until, having loaded his ship with plundered store, he turned his prow homeward. When off the coast of his native country, a thick mist arose, so that the pilot knew not which way to steer. The wind blew a gale all day, but sank into silence at eventide; and Sir Ralph stood upon the deck, vainly looking out into the dark to catch sight of land.

> " Quoth he, 'It will be brighter soon,
> For there's the dawn of the rising moon.'

> " ' Canst hear,' said one, 'the breakers roar?
> For yonder, methinks, should be the shore.'
> ' Now where we are I cannot tell—
> I wish we heard the Inchcape Bell !'

> " They hear no sound—the swell is strong ;
> Though the wind hath fallen they drift along,
> Till the vessel strikes with a shivering shock—
> ' O Heaven ! it is the Inchcape Rock !'

> " Sir Ralph the Rover tore his hair,
> And cursed himself in his despair.
> The waves rush in on every side ;
> The ship sinks fast beneath the tide !

> " Down, down they sink in watery graves,
> The masts are hid beneath the waves !
> Sir Ralph, while waters rush around,
> Hears still an awful, dismal sound,—

> " For even in his dying fear
> That dreadful sound assails his ear,
> As if below, with the Inchcape Bell,
> The devil rang his funeral knell."

When Mr. Robert Stevenson made his first landing on the rock in 1800, he discovered in almost every chink and cranny painful proofs of the sad disasters of which it had been the cause,—such as bayonets, musket-balls, and innumerable fragments of iron. All the more perishable materials had been swept away; and a silver shot-buckle was the only vestige of wearing-apparel to speak of the loss of many who had here met their unexpected doom. Nor was it simply on the unbeaconed and unlighted rock that ships were destroyed; not a few were cast on the neighbouring shores, from the anxiety of their pilots to avoid the dreaded danger. Mr. Stevenson records a melancholy example. During a three days' gale in 1799 a large fleet of vessels were driven from their moorings in the Downs and Yarmouth Roads, and from their southward courses. Borne north by the fury of the blast, these ships might easily have reached the anchorage of the Firth of Forth, for which the wind was favourable; but night came on, and fearing the Bell Rock, their ill-fated steersmen resolved to keep at sea, but drifting before a pitiless storm on a dark December night, they lost their reckoning and were hopelessly wrecked, two of them on the Bell Rock, and about seventy on the eastern shores of Scotland, where, alas! many of their brave crews perished. This fatal catastrophe, says Mr.

Stevenson, is the more to be lamented when we consider that a light upon the Bell Rock, by opening a way to a place of safety, would infallibly have been the means of preventing it. And that this opinion was justifiable we know from the fact that not a single ship has been lost on the rock since the lighthouse was completed—three quarters of a century ago.

It was not until 1786 that a Lighthouse Board for Scotland was established. At that time the chief lights on the Scottish coast were the chauffer and coal-fire, already spoken of, on the Isle of May, in the estuary of the Forth, and a similar chauffer on the Little Cumbrae, in the estuary of the Clyde. But the dangerous character of the Bell Rock soon attracted the attention of the Commissioners,* and they began to contemplate the erection of a lighthouse upon it. In 1806 they obtained an Act of Parliament authorizing them to proceed with it; and in the following year operations were begun under the superintendence of Mr. Robert Stevenson, who was chosen to carry out the design and plans furnished by the celebrated engineer, Mr. (afterwards Sir John) Rennie. The execution of the work, attended as it was with exceptional difficulties, occupied about four years, and the outlay involved amounted to £61,331, 9s. 2d., toward which Government advanced a sum of £30,000.

The 17th of August 1807 was the first day. The

* As early as 1793 Admiral Sir A. Cochrane had publicly pointed out its danger.

erection of a workshop and of a temporary residence for the workmen occupied the whole season, the supports having to be firmly secured in holes dug out of the solid rock. The hardness and compactness of the sandstone soon blunted their tools, and a smith with his forge was kept in constant employment. But it often happened, says Mr. Stevenson, in his highly interesting "Account of the Bell Rock Lighthouse," "to our annoyance and disappointment, that when the smith was in the middle of 'a favourable heat,' and fashioning some useful article, or sharpening the tools, after the flood-tide had compelled the men to strike work, a sea would come rolling over the rock, dash out the fire, and endanger that indispensable implement, the bellows; or, if the sea were smooth, while the smith often stood at work knee-deep in the water, the tide rose imperceptibly, first cooling the exterior of the fire-place or hearth, and then quietly blackening and extinguishing the fire from below." Mr. Stevenson was frequently amused at the anxiety and perplexity of the unfortunate smith when coaxing his fire, and endeavouring in vain to contend against the rising tide. Obviously the work could make but slow progress, until the work-ship, which was also intended to serve as a beacon, was completed, and the smith protected against the insidious waters.

But something more serious occurred on the 2nd of September. The first cargo of stones had been landed, and two-and-thirty men were engaged in various branches of work, when a gale arose, and the tender or store-ship, the *Smeaton*, which lay-to by the rock, broke adrift from her moorings. This mishap was known at first only to Mr. Stevenson and his "landing-master," who fully appreciated the gravity of the situation: thirty-two men on an isolated rock, which at flood-tide had twelve feet of water upon it, with only two boats at hand, and these, in foul weather, not capable of carrying more than eight men each.

While the men were at work excavating the rock and boring with their tools, and while the din of their hammers and the clang of the smith's forge filled the air, the life and movement of the scene kept Stevenson from fully realizing the threatening possibilities of danger. But gradually the tide began to rise, and slowly it gained upon those who were engaged on the lower portions of the sites of the beacon and light-house. From the run of the sea upon the rock, the forge-fire was more quickly extinguished than usual; and the smoke subsiding, every object around was distinctly revealed. The majority of the men made to-wards their respective boats to get their jackets and stockings, when, to their astonishment, they found only two boats on the rock, the third being adrift with the *Smeaton*. Not a word was uttered, but all appeared to be silently counting their numbers, while gazing from one to another with expressive looks of perplexity and dismay.

Meanwhile, Mr. Stevenson was anxiously considering

what expedients might be adopted for the safety of the party. The most feasible plan seemed to be, that as soon as the waves reached the highest point of the rock, the men should strip off their upper garments, and, while a certain number got into each boat, the remainder should support themselves by the gunwales, and the boats row gently towards the *Smeaton.* Stevenson would fain have proposed this plan, but on attempting to speak he found his mouth so parched that he could not articulate a syllable. Turning to one of the rock-pools, he sipped a little water, and obtained immediate relief. Great was his joy when, on rising from this unpleasant beverage, he heard a voice exclaim, "A boat! a boat!" and, on looking around, perceived a large lugger approaching, which proved to be a pilot-boat from Arbroath, express with letters. The crew willingly took on board Mr. Stevenson and his companions, and delivered them from their unpleasant position.

On the 6th of September, when the whole company were on board the *Pharos,* or light-ship, a tremendous gale arose, which prevented them from approaching the rock for two days.

About two o'clock P.M., says Mr. Stevenson, a very heavy sea struck the ship, flooded the deck, and poured into the berths below. All on board thought she had foundered, and that their last moment had come. Below deck total darkness prevailed; several of the workmen were praying, repeating hymns, or uttering devout ejaculations; others protested that **if they should fortunately** be spared **to** reach land **once more,** nothing would induce **them** to tempt the treacherous waves again. In the confusion Stevenson made **his** way upon deck. The spectacle which met his gaze was not very encouraging. The watch on deck stood lashed to the foremast that he might not be washed overboard. The billows appeared to be ten or fifteen feet in height **of** unbroken water, and seemed to threaten with destruction **the** small craft that **laboured among them.** The horizon was heavy with clouds, and through the **gathering** mist it was impossible to discern **the** Scottish shore.

Happily, the gale abated about **six** o'clock **in the** evening, and next morning the men rose under a comparatively **serene** sky. The waves still rolled **very heavily, and at** the Bell Rock hurled up **their** spray in columns forty to fifty **feet** high. As soon as Mr. Stevenson was able to land, **he** found abundant evidence of their force: six large blocks of granite had been heaved from their **places and** rolled over a rising ledge into a hole some twelve **or** fifteen paces distant. The ash-pan of the smith's forge, **with its heavy cast-**iron back, had also been wrested from its position, the connecting chain broken, and the ponderous **articles** swept right over to the opposite **side of the** rock.

By the latter end of October, however, the initial stage **of** the undertaking **was** completed, and **a beacon** erected, **which** consisted **of twelve beams of timber** forming **a common** base **of thirty-six feet, with fifty**

feet of height, the whole being securely fastened to the
rock by batts and chains of iron. The third and topmost compartment of this beacon was used as a barrack
for the men while the work was in progress; on the
second floor, twenty-five feet from the rock, the mortar
was prepared, and the smith toiled away at his forge.
On several occasions the violence of the sea lifted this
floor, but none of the batts were shaken, and it remained on the rock until removed in the summer of
1812.

The operations of the second season began at the
earliest date the weather permitted. A second forge
having been set up on the beacon, the smiths plied
their trade with laudable activity,—sharpening picks
and irons for the masons, and manufacturing batts,
movable cranes, and other apparatus for use on the
railways. The landing-master's crew were occupied in
helping the millwrights to lay down the cast-iron railway for the conveyance of materials. Seamen are no-
torious for their dexterity at all kinds of manual work,
and now they handled promptly and freely the boring
irons, and assisted in all the railway operations, acting
by turns as boatmen, carpenters, and smiths. Both on
the rock and on shipboard they were the inseparable
companions of every work connected with the erection
of the Bell Rock Lighthouse.

It might naturally be supposed that a score and
more of masons occupied with their picks in executing
and preparing the foundation of the lighthouse would,

in the course of a three hours' tide, make a considerable impression upon an area of even forty-two feet in diameter. But in proportion as the foundation was deepened, the rock was found to be much harder and more difficult to work, while the baling and pumping of water became more troublesome. A joiner was kept almost constantly employed in fitting the picks to their handles, which, as well as the points of the irons, were very frequently broken. At eight o'clock the water overflowed the site of the building, and the boats left the rock with all hands for breakfast.

The Bell Rock at this time presented a very striking appearance, with impressive contrasts of light and shade which would have delighted the eye of Rembrandt. Its narrow surface was crowded with busy men, who were in constant movement. The two forges flamed, one above the other, like Cyclopean furnaces; while the anvils thundered with the rebounding clash of their wooden supports as if in rivalry with the regular roll and rush of the ocean tide. During the night, if the men were at work, a still more picturesque scene might have been witnessed by any passing vessel. Even to the operatives themselves the effect of extinguishing the torches was sometimes startling, and made darker, as it were, the darkness of the night; while the sea, lighted up by a phosphorescent glow, rolled in upon the rock its long heaving waves of fire.

Sufficient progress having been made in the work of preparation, it was resolved that the foundation-stone should be laid on the 10th of July. The ceremony attending it was very simple, as needs must be on a rock less than a fifth of a mile long. Mr. Stevenson, supported by his three assistants, applied the square, the level, and the mallet in due form, and pronounced the benediction, "May the great Architect of the universe complete and bless this building!" The men gave three hearty cheers—assembled there on the lone, weedy rock, with a gray waste of waters all around—and then drank success to the future operations with the greatest enthusiasm.

The first course of masonry was rapidly laid down. It was only one foot in thickness, but it contained 508 cubic feet of granite in outward casing; 8,076 cubic feet of Mylnefield stone in the hearting; 104 tons of solid contents; 132 superficial feet of hewing in the face-work; 4,519 superficial feet of hewing in the beds, joints, and joggles; 420 lineal feet boring of trenail holes; 378 feet lineal cutting for wedges; 246 oaken trenails; and 378 oak wedges in pairs. These figures may not be interesting in themselves, but will serve, perhaps, to awaken in the reader's mind some idea of the colossal amount of labour involved in the construction of a lighthouse of the first class.

By the end of the season the masonry was raised to a level with the highest part of the margin of the foundation-pit, or about five and a half feet above the lower bed of the foundation-stone. Work was discontinued on the 21st of September.

During the **winter** a stock of materials having been obtained **from** the granite quarries of Aberdeenshire, **for an external casing to the height of thirty** feet, **and from the freestone** quarries of Mylnefield, **near** Dundee, **for the inside** and upper walls, a number of masons **were kept employed in the** work-yard **at Arbroath,** and every preparation made for resuming work on the rock in the following spring. The **stones were** wrought with great accuracy—laid **upon a** platform course by **course,** numbered and **marked as they were each to lie in the building, and then set aside for transportation, everything being done** with **wonderful order** and **despatch. On the rock,** operations **were recommenced on** the 27th **of May; and, despite various accidents and delays, and considerable** obstruction **from inclement weather, had progressed** so far by **the end of June that the** workmen **were able to continue their** labours on the masonry while **the rock was** under **water. On** the 8th **of July it** was observed, **to the general** satisfaction, that the **tide for the** first time **failed to** cover **the** masonry **at high water.** Flags were accordingly **hung out at every vantage point, as well as on** board **the** yacht, **the tender, the** stone-**praams, and** the floating-light; **a salute of three guns was fired;** and the loudest and heartiest cheers imaginable **rent the air and** mingled **with** the myriad voices of **the** waters. By **the 25th of** August, **when** the work **was** discontinued **for the autumn and** winter months, the solid **portions of the Bell** Rock Lighthouse had been raised thirty-one and a half feet above the rock, and seventeen feet above high water at spring tides.

Having in a couple of seasons landed and built up more than one thousand four hundred tons of stone,

THE BUILDING OF THE BELL ROCK LIGHTHOUSE.

while the work was low down in the water, and as not more than seven hundred tons were required to complete the masonry, Mr. Stevenson allowed himself to anticipate that another season would complete his undertaking. But success absolutely depending on the stability of the beacon, he paid frequent visits to the rock in the course of the winter to make sure that it defied the fury of winds and waves.

The work of the fourth and last season was begun on the 10th of May. The artificers took possession of the beacon, which consisted at this time of three floors —one occupied as the cook-house and provision-store; the second, divided into two cabins, for the engineer and foreman respectively; and the third, provided with three rows or tiers of sleeping-berths, to accommodate about thirty men. Below these three floors was a temporary floor, at the height of twenty-five feet above the rock, used as the smith's workshop—only one was now required—and for preparing mortar. A wooden bridge connected the beacon with the growing lighthouse.

The room or cabin occupied by Mr. Stevenson he graphically describes. It measured, he says, not more than four feet three inches in breadth on the floor; and though, from the oblique direction of the beams of the building, it widened upwards, yet was he unable when standing up to stretch out his arms to the full extent; while its length barely admitted of a hammock being slung for the night. His folding writing-table was attached with hinges immediately under the small window of the apartment; and his boots, barometer, thermometer, portmanteau, and two or three camp-stools, formed the bulk of his movables. His diet being plain, the table-equipage (as Lord Lytton somewhere calls it) was correspondingly simple, though there was a general air of comfort and neatness, the walls being covered with green cloth, formed into panels with red tape, and his bed festooned with curtains of yellow cotton-stuff.

We have not the space, and probably our readers would not have the inclination, to dwell upon the daily details of Mr. Stevenson's great undertaking. We pass on to the 29th of July as one of its epoch-days, for then the *last stone* was landed on the Bell Rock. You may be sure that the occasion was duly celebrated. On the 30th, the last and ninetieth course was laid, completing the outer wall of the lighthouse tower, and its designer then solemnly uttered a suitable benediction: "May the great Architect of the universe, under whose blessing this perilous work has prospered, preserve it as a guide to the mariner!"

On the 17th of December 1810, an advertisement in the public journals made known to the public that Mr. Stevenson's great work had been completed, and that thenceforth the perils of the Bell Rock would virtually cease to exist for the mariner:—"A lighthouse having been erected upon the Inch Cape or Bell Rock, situated at the entrance of the Firths of Forth and Tay, in north lat. 56° 29′, and west long. 2° 22′, the Commissioners of the Northern Lighthouses hereby give notice that the light will be from oil, with reflectors, placed at the height of about one hundred and eight feet above the medium level of the sea. The light will be exhibited on the night of Friday, the first day of February 1811, and each night thereafter, from the going away of daylight in the evening until the return of daylight in the morning. To distinguish this light from others on the coast, it is made to revolve horizontally, and to exhibit a bright light of the natural appearance and a red-coloured light alternately, both respectively attaining their greatest strength, or most luminous effect, in the space of every four minutes: during that period the bright light will, to a distant observer, appear like a star of the first magnitude, which, after attaining its full strength, is gradually eclipsed to total darkness, and is succeeded by the red-coloured light, which in like manner increases to full strength, and again diminishes and disappears. The coloured light, however, being less powerful, may not be seen for a time after the bright light is first observed. During the continuance of foggy weather and showers of snow, a bell will be tolled by machinery, night and day, at intervals of half a minute."

The Bell Rock Lighthouse now exhibits a white and red light alternately, revolving at intervals of a minute. The building is of a circular form, measuring forty-two feet in diameter at the base, but diminishing or tapering towards the top, so that it measures only thirteen feet at the neck, immediately below the light-room. Its total height is one hundred and fifteen feet. To the height of thirty feet it is entirely solid, with the exception of a "drop-hole" of ten inches in diameter for the weight of the machinery which moves the reflectors. The ascent to the doorway, which is placed immediately above this solid substructure, is by a ladder of gun-metal. A narrow passage leads from the door to the

THE BELL ROCK LIGHTHOUSE.

staircase, where the walls are seven feet thick; from the top of the staircase, which is thirteen feet high, the walls gradually diminish up to the summit. Above the staircase, access to the different apartments is furnished by wooden ladders, the remaining fifty-seven feet of masonry being divided by five stone floors into six rooms for the light-keepers and stores. The three lower rooms—for coals, water, and oil respectively—have each two small windows, while the three upper rooms—the bedroom, the kitchen, and the sitting-room, which is adorned by a marble bust of Stevenson the engineer—have each four; and all are provided with strong shutters to defend the glass panes in stormy weather. The two first courses of the masonry are entirely sunk into the rock; in every course the stones are dovetailed and inserted into each other, so that the building forms one closely connected mass, which seems as if it would last as long as time itself.

The light-room, which is constructed of cast iron, and glazed with polished glass, is octagonal in shape, twelve feet in diameter, and fifteen feet in height. It is covered with a dome, and terminates in a ball.

The manner in which this noble structure bears the onset of the waters has been described by Mr. Stevenson. During the gales of winter, and when viewed from the Forfarshire coast, it appears in a remarkably interesting aspect, standing proudly among the waves, while the sea around it is in the wildest state of agitation. The light-keepers do not seem to be in motion, but the scene is by no means still, for the clang and clamour, the motion and fury of the billows, are incessant. The seas mount to the height of about seventy feet above the rock, and after expending their force in a perpendicular direction, fall in masses of foam round the base of the lighthouse, while considerable portions of the spray seem to adhere, as it were, to the building, and collect on its sides in snow-white froth. Some of the great waves burst and are expended upon the rock before they reach the lighthouse; while others strike the base, and embracing the walls, meet on the western side, where the violent collision churns the eddying waters into the wildest foam.

The management of the lighthouse is organized as follows:—At Arbroath, about eleven miles distant, is stationed a cutter, which, once a fortnight, or in the course of each set of spring-tides, visits the rock, in order to relieve the light-keepers and replenish their stock of fuel and provisions. There are four light-keepers, three of whom are always on duty, while one is ashore. If the weather be favourable, each light-keeper is six weeks on the rock, and a fortnight on land with his family. The wage is from £50 to £60 a year, with a stated allowance of bread, beef, butter, oatmeal, vegetables, and small beer, and fourpence a day extra for tea. A suit of uniform is also provided once in three years.

The watches in the light-room are relieved with as much punctuality as on board a man-of-war—no keeper being allowed to quit his station until the relief appears, on pain of immediate dismissal. To insure the strictest regularity in this respect, a timepiece is placed in the light-room, and bells are hung in the bedrooms of the dwelling-houses, which, being connected with the lighthouse by mechanical appliances, can be rung as necessity requires.

At Arbroath, as at other stations, the light-keepers' dwellings are very neatly built and comfortably arranged, each with its little garden attached. There are also suitable storehouses, a room for the master and crew of the lighthouse tender, and a signal-tower fifty feet high, on the summit of which is a small observatory, with an excellent achromatic telescope, a flag-staff, and a copper signal-ball measuring eighteen feet in diameter. A similar ball crowns the lighthouse dome, and by this means daily signals are exchanged to signify that all is well. Should the ball at the top fail to be raised, as is the case when particular supplies are needed, or either of the light-keepers has been stricken with illness, the tender is immediately despatched with assistance.

Two bells are suspended to the lighthouse balcony, and in stormy weather are rung every half minute by the mechanical action of the lighting apparatus.

A curious accident occurred here about ten o'clock on the night of the 9th of February 1832. A large herring-gull flew against one of the south-eastern mullions of the light-room with so much violence that two of the polished glass plates, which each measure about two feet square and a quarter of an inch thick, were dashed to atoms and scattered over the floor, to the great alarm of the keeper on watch, and his two associates, who rushed instantly into the light-room. It happened, fortunately, that though one of the red-shaded sides of the reflector-frame was revolving at the moment, the fragments were so minute that no injury was done to the valuable red glass. The bird was found to measure five feet from tip to tip of its extended wings. A large herring was in its gullet, and in its throat a piece of plate glass about an inch in length.

Thrushes and blackbirds are occasionally killed in winter by dashing against the lantern. The keepers catch a few fish.

At the entrance to Montrose Harbour a white inter-mittent light has been shown from *Scurdy* (or *Montrose*) *Ness*, since February 1870. The want of such a beacon and sea-mark had long been felt by the seafaring com-munity. The iron-bound shore between the Bell Rock and the Girdleness, a stretch of fifty miles, is perhaps one of the most dangerous parts of the east coast of Scotland, and has been the scene of numerous ship-wrecks and great loss of life. At no point within these limits had so many disasters occurred as at the entrance to Montrose Harbour, now fortunately pro-tected by the Ness light. Bounded on the one side by large outlying, and in some instances hidden, rocks, and on the other by leagues of sandy shore, whilst the channel itself is exceedingly narrow, the entrance to Montrose Harbour is very difficult for navigation, and especially is it hazardous in stormy weather.

Again, on the north side, and within a very short distance of the new lighthouse, lies the Annat, a sand-bank on which many vessels have been wrecked in attempting to make the harbour. Situated at the Point, on the southern side of the channel, the lighthouse commands a fair-weather range of seventeen nautical miles. It stands on solid rock; its foundation is of granite; and the shaft or tower of brick is painted white. From base to vane the measurement is one hundred and twenty - seven feet; at the base, twenty-three feet two inches in diameter; at the top, sixteen feet. A spiral staircase of about one hundred and forty steps leads to the top of the tower; the ascent to the light-room and lantern is continued by ladders.

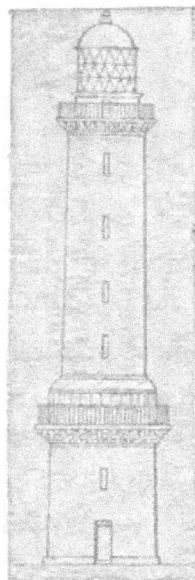

GIRDLENESS LIGHT-HOUSE.

Passing Stonehaven, we come to *Girdleness*, on the Aberdeen coast, in lat. 57° 8′ 33″, where a well-built and handsome tower of stone, with double walls, one hundred and twenty feet high, was erected by Mr. Robert Stevenson in 1833, at a cost of £12,940, 5s. 1d. It shows two fixed white lights, one seventy feet above the other. In November 1858 the light on one occasion was suffered to go out, the keeper falling asleep. He was dismissed for his breach of duty.

At *Buchanness*, beyond Port Errol, a light-tower (of

stone, with double walls), one hundred and fifteen feet high, was erected by Mr. R. Stevenson in 1827, at a cost of £11,912, 5s. 6d. As many as fifty-five foggy days in one year have been noted at this station. The light here is white, with a five-seconds flash, and visible seventeen miles.

The lighthouse on *Kinnaird Head*, the turning-point of the Moray Firth, originally converted from an old castle by Thomas Smith, in 1787, was restored in 1851. It has a fixed white light, with red sector. *Covesea Skerries* lighthouse, on the shore of the firth, was erected by Alan Stevenson in 1846, at a cost of £11,514, 16s.; is one hundred and eighteen feet in height, built of stone, and equipped with a revolving light (white and red). At *Chanonry Point*, the lighthouse, erected by Alan Stevenson in 1546, as a guide to open up Inverness Firth, is built of stone, and forty-two feet in height. Cost £3,571, 17s. 2d. Has a fixed white light, visible eleven miles.

On *Cromarty Point* may be noticed a lighthouse, by Alan Stevenson, exactly similar to the one just described, and first lighted in the same year. Cost £3,203, 9s.

Passing the great fishery-centre of Wick, we come in sight of *Noss Head*, on the shore of Caithness, in lat. 58° 28′ 38″. The stone tower here, sixty-eight feet high, was built by Mr. Alan Stevenson, in 1849, at a cost of £12,149, 15s. 8d. It displays a revolving white light, with red sector, at half-minute intervals.

The rocky islets of the *Pentland Skerries* lie off the north-east coast of Scotland, in lat. 58° 41′ 22″. In 1793 the Liverpool ship-owners petitioned the Trinity House to erect a light upon them, to assist vessels navigating the Pentland Firth. In the following year two towers were erected by Mr. Thomas Smith, at a distance from each other of a hundred feet, which, in 1833, were rebuilt by Mr. Thomas Stevenson. They are of stone, and respectively one hundred and eighteen and eighty-eight feet in height from base to vane. Both display fixed white lights, visible nineteen and eighteen miles.

At *Dunnet Head* the Pentland Firth attains its maximum breadth (about twelve miles). The Head is an extensive promontory stretching into the firth, and forming the eastward boundary of Thurso Bay. It consists of several hills broken up with valleys, but presenting seaward a boldly irregular front, varying from one hundred to four hundred feet in height. It is joined to the mainland by a narrow isthmus, about a mile and a half broad. A stone lighthouse, about sixty-six feet high, was erected by Mr. Robert Stevenson here in 1831, which exhibits a fixed white light, visible twenty-four miles. The centre of the lantern is three hundred and forty-six feet above high-water mark. Cost of lighthouse, £9,135, 15s. 3d.

The low western boundary of Thurso Bay is called *Holburn Head*, in lat. 58° 36′ 55″, where a light-tower, fifty-five feet high, was erected in 1862. It shows a white light, flashing every ten seconds, with red sector.

The shores of the Orkneys, celebrated by **Sir Walter Scott** in his romance of "The Pirate," are well lighted: on *North Ronaldshay*, lat. 59° 23′ 15″, on the one hand, and *Cantick Head*, South Walls Island, lat. 58° 47′ 18″, on the other; and between these extreme points, *Gremsay Isle*, in Hoy Sound, two lights; *Auskerry*, Stronsay Firth, and *Start Point*, Sanday Isle.

A lighthouse was erected by Mr. Thomas Smith on *North Ronaldshay*, the northern point of the Orkneys, in 1789; but after an experience of twelve years the position was thought to have been unfortunately selected; and it was then determined to plant a new one on Start Point, Sanday, and convert the North Ronaldshay lighthouse into a beacon. This change, however, was not welcomed by mariners, and in 1854 the tower was rebuilt by Mr. Alan Stevenson and a light again set up at North Ronaldshay. The building is of brick, has double walls, and measures one hundred and thirty-nine feet in height; cost £12,927, 19s. 4d.; and exhibits a white light, flashing every ten seconds, visible eighteen miles.

In 1796 three homeward-bound vessels were lost in these waters, and eight more in the next three years. For the better protection of navigation it was then resolved (1802) that a beacon, or solid tower of masonry, should be raised upon *Start Point,* the eastern extremity of Sanday Island; and in 1802, Mr. Robert Stevenson, the engineer, began the necessary operations. The foundation stone was laid on the 13th of May, and in the following September the beacon was completed, measuring ninety-one feet in height, and terminating with a solid ball of masonry fifteen feet in circumference. It soon appeared, however, that this was insufficient to prevent the occurrence of wrecks in almost as great number as before. It was proverbial with the inhabitants to observe that, if wrecks were to happen, they might as well be sent to the poor island of Sanday as anywhere else. In fact the islanders, both here and in the archipelago generally, had long lived upon the proceeds of the wreckage—that melancholy harvest of the sea—and the remains of many a "tall ship" had been to them a source of sustenance and profit. For example, though quarries are here very numerous, and the stone specially adapted for the construction of dikes, yet it was not uncommon to find enclosures fenced round, even when of considerable extent, with ship-timbers. A "park" (*Anglicè,* field) might be seen enclosed with palings of mahogany and cedar-wood from the wreck of a Honduras-built ship. In one island it is recorded that a ship laden with wine having been driven on its shores, the inhabitants took claret to their barley-meal porridge, instead of their usual beverage. When Mr. Stevenson complained to his pilot of the inferior quality of his boat's sails, he replied, with grim humour, "Had it been God's will that you came na here wi' these lights, we might a' had better sails to our boats, and more o' other things." A higher rent was given for farms on the coast than they were really worth, in consideration of the gain that would probably arise from shipwrecks.

Ultimately the beacon at Start Point was converted by Mr. Robert Stevenson into a lighthouse, which, on the 1st of January 1806, was illuminated for the first time. It shows a fixed red light, visible in fair weather at a distance of fourteen miles.

On the small, uninhabited island of *Auskerry*, in Stronsay Firth, in lat. 59° 1' 25", a lighthouse was erected in 1867. It is built of white brick, and is one hundred and twelve feet high. The light is white, fixed, and visible sixteen miles.

Hoy Sound is protected by two lighthouses; one, a tower one hundred and eight feet high, on the north-east point of *Gremsay* (or Græmsay); the other, two thousand two hundred and thirty-seven yards distant, on the north-west point of the same island, thirty-eight feet high. The former is a red light, fixed, with white sectors, visible fifteen miles; the latter, a white light, fixed, visible twelve miles. Date, 1851. Both are built of stone, and were designed by Mr. Alan Stevenson. Cost £15,880, 19s. 7d.

At *Cantick Head*, on South Walls Island, lat. 58° 47' 18", a light-tower of brick, seventy-three feet high, was erected in 1838.

We now come to the Shetland group.

At *Sumburgh Head*, lat. 59° 51′ 15″, distant about two hours' sail from Lerwick, and elevated three hundred feet above the sea-level, stands one of Robert Stevenson's admirably constructed lighthouses, solid and yet graceful, with its lines telling of beauty as well as strength. It was built of stone, with double walls, in 1820–1, at a cost of £10,087, 1s. 11d., and is fifty-five feet high; its white light, fixed, at an elevation of three hundred feet, has a fair-weather range of twenty-four miles.

Sumburgh Head is a bold and rocky precipice on the coast of Fair Island, near the "Fitful Head" described by Scott in his "Pirate;" and between this island and the mainland boils and rages the famous tideway of "the Roost of Sumburgh," almost impassable in tempestuous weather. Fatal and frequent wrecks took place here prior to the erection of the lighthouse.

Our next point, on the east side of the entrance to Lerwick, is the island of *Bressay*, separated from the mainland by Bressay Sound—a great rendezvous of the English and Dutch herring-boats, and a kind of natural haven, which has often afforded shelter to our men-of-war. The lighthouse here is of white brick, fifty-three feet high, and was erected in 1858 by Messrs. Stevenson. It shows a revolving light, red and white alternately, at a minute's interval; visible sixteen miles. Cost £5,163, 7s. 6d.

On the *Out Skerries*, about twelve miles north of Lerwick, in lat. 60° 25′ 24″, a stone tower, ninety-eight feet in height, was erected in 1856–8 from Thomas Stevenson's designs. It exhibits a white light, revolving at intervals of a minute, and visible eighteen miles. Cost £21,450, 18s. 2d.

OUT SKERRIES LIGHTHOUSE.

Lastly, on North Unst, the northernmost island of the Shetland group and of Great Britain, in lat. 60° 31′ 59″, we find the *Muckle Flugga Lighthouse*. It is built of brick, from Messrs. Stevenson's designs, at a cost of £32,478, 15s. 5d., and situated upon an outlying rock of conical form, called a "stack," nearly two hundred feet above the sea-level, and on the north opposes an

almost precipitous front to the surging seas. On the
south its declivities are less abrupt, but scarcely easier
of access; and its summit is only wide enough to receive
the foundation of the light-tower. This is a handsome
structure of stone, fifty-four feet high, erected in 1854–8,

NORTH UNST LIGHTHOUSE.

at a cost of £32,478, and contains, besides the lantern-
room, a sleeping-chamber, kitchen, and store-room. At
its base is a semi-circular building, used for supplies of
oil, charcoal, and fresh water. The keepers' residences
are situated on the island of Unst, four miles distant.

Cape Wrath, the north-west turning-point of Scotland, in lat. 58° 37′ 30″ and long. 4° 59′ 41″, is a bold and rugged headland, defying the heaving waters of the Atlantic with rocky steeps and swart perpendicular cliffs that attain an elevation of three hundred and three hundred and fifty feet. At their feet the currents swirl dangerously over a sunken reef, and the surface of the sea is sprinkled with desolate islets, varying in size and character, but all repellent and sterile. Welcome in such a scene is the friendly radiance of the lighthouse—a weather-beaten tower of granite, erected by Mr. Robert Stevenson in 1838, at a cost of £13,550, 18s. 9d. It exhibits alternately a brilliant red and white light, revolving at a minute's interval, and visible for twenty-seven miles.

We have not left it far behind us before we come within range of the intermittent flashes of the light upon *Ru Stoer*, or South Ear of Ru Stoer, in lat. 58° 14′ 10″. The tower, of white stone, was erected from Mr. Thomas Stevenson's designs in 1870, and is forty-seven feet high. It carries a white light, visible for sixty seconds, then dark for thirty. The focal plane is one hundred and ninety-five feet above high-water mark.

Our cruise now brings us to the west coast, and we sail in the shadow of the island of *South Rona*, at the end of which a brick tower, forty-two feet high, with keepers' dwellings, was erected by Messrs. Stevenson in 1857. It exhibits a white light, flashing every twelve seconds, and visible twenty-one miles. At *Kyleakin*—a name which reminds us of Haco of Norway and his maritime exploits—on the shore of Loch Alsh, and in lat. 57° 16′ 59″, was erected—also in 1857, and by Messrs. Stevenson—a brick lighthouse, seventy feet high, showing a fixed white light, with red sector, visible twelve miles. And in November ·1857, the Messrs. Stevenson erected a brick tower, sixty-three feet high, on *Oronsay Island*, in Sleat Sound, which throws a fixed white beam of light over twelve miles of sea. Cost of South Rona, £5,063, 4s. 10d.; of Kyleakin, £6,210, 19s.; of Oronsay, £4,527, 17s. 10d., which is complete in every detail.

Stretching across to the Hebrides, we find them provided with six lighthouses, besides harbour and pier lights, for the better guidance of the navigators of their stormy waters. On the north point of the *Butt of Lewis*—Lewis is the northern portion of Long Island, and in English fiction will not fail to be remembered as the home of Mr. William Black's enchanting "Princess of Thule"—a stately lighthouse tower, one hundred and twenty feet high, was erected in 1852. Its fixed white light can be seen at a distance of nineteen miles. On *Arnish Point*, opening up the approach to Stornoway, an iron tower, forty-five feet high, was erected in 1852, from Mr. Alan Stevenson's designs. It shows a white light, revolving at half-minute intervals; while a ray from one of its lower windows is, by an ingenious arrangement, made to illuminate a glass prism in a lantern attached to the top of a beacon on *Arnish Reef,* some distance out in the bay, and thus to indicate more clearly the proper channel. The effect is so good that for a long time the Lewis fishermen insisted that there was a real light on the beacon. Cost of lighthouse, £6,380, 19s. 5d.

ARNISH LIGHTHOUSE AND BEACON

At *Monach,* off the west coast of North Uist, on Shillay Island, there is a fine light, flashing at intervals of ten seconds, from a tower one hundred and thirty-three feet high, erected in 1864 from Mr. Stevenson's designs. Visible eighteen miles.

There is another at *Scalpay,* Glass Island, lat. 57 51' 25". The building, one hundred feet in height, designed by Mr. T. Smith, dates from 1789. The light is white, fixed, and visible twelve miles.

At *Ushenish,* on the east side of South Uist, the

tower is of brick, only thirty-nine feet high. It was erected in 1857 by Messrs. Stevenson, and in 1885 furnished with a white occulting light, visible eight seconds and eclipsed for sixteen, which can be sighted at a distance of eighteen miles. Cost £8,809, 4s. 3d.

There was once upon a time a Highland chieftain, a M'Neill of Barra, who, after finishing his daily banquet of cod, ling, and cockles, was wont to ascend to the top of his castle of Chisamil, and in sonorous Gaelic proclaim :—"Hear, O ye people ! and listen, O ye nations ! The great M'Neill of Barra having finished his dinner, all the princes of the earth are at liberty to dine !" Great quantities of cod and other fish are still caught off the coast of Barra, and for the convenience of the fishermen, a lighthouse was erected in 1833 on the southernmost island of the Barra group—Bernera or Long Island. The cliffs here are exceedingly various in outline : inclining, perpendicular, projecting ; some overhang the waters with beetling brows, some are deeply fissured and broken up into irregular precipices ; but all, in the summer months, resound with the discordant cries of kittiwakes, guillemots, auks, and puffins. On *Barra Head*, one of these formidable crags, is perched the solid stone tower of the lighthouse, sixty feet in height, with its lantern six hundred and eighty-three feet above high-water mark. Designed by Mr. Robert Stevenson ; cost £13,087, 13s. 11d. Some twenty years ago it was visited by the late Dr. William Chambers, as one of the Commissioners of

the Northern Lights, and he has given a very interesting account of what he saw.

With its surrounding walls and gates, the lighthouse
establishment has, he says, somewhat the aspect of a
fortification. "The whole of the buildings are of a
beautiful white granite, quarried in the island......An
interior paved court is environed by the houses of
three keepers; and passing them, we reach the tower
for the light, with its winding stair, which all immediately ascend. What an outlook from the upper
story down to the sea, which surges seven hundred
feet below! and what myriads of sea-birds screaming
and fluttering on ledges of this tremendous precipice!
I have seen it stated that these cliffs excel in grandeur
anything of the kind in the Hebrides, and can scarcely
doubt that such is the case. On a projecting point
immediately in front of the lighthouse are the ruins
of an old castle or keep, once the stronghold of some
Hebridean chief. As usual, before departure, we visited
the several houses of the keepers, and in one of them
some information was picked up respecting a water-
mill which had excited our curiosity. This mill is
entirely the handiwork of an ingenious assistant lighthouse keeper (a Fife man), who diverted his leisure
hours in its construction. He erected the building,
covered it with a tarpaulin roof, and fabricated the
whole of the grinding apparatus. The most difficult
part of the undertaking was accomplished by adapting
an old cart-wheel. The idea of erecting a mill was

suggested by the absence from the island of all means for grinding except by a primitive species of hand-querns. It turned out to be a grand conception, this mill. Glad of the opportunity of so easily transforming their corn into meal, the crofters besought the privilege of using it, which was of course allowed ; and as money happens to be a rare article in Bernera, the multure was arranged on the convenient footing of giving a lamb for a grist, be the quantity much or little."

The next lighthouse to which our rapid survey brings us is that of SKERRYVORE (nine miles from Tiree Island), which, in lighthouse history, occupies a position of honour like that of the Eddystone, of the Bishop Rock, or of the Bell Rock. It resembles them in its isolation—a lonely tower of stone, erect amid the gray waste of waters; it resembles them in the beauty of its form and the harmony of its proportions. The loftiest light-tower in the world, with the exception of the *Tour de Cordouan*, it is remarkable for the simplicity of its structure, and its combination of all that science requires of strength and all that art requires of grace.

The Skerryvore reef is, in its main features, a counterpart of the Inch Cape or Bell Rock. Curiously enough, it lies in the same parallel of latitude (56° 19′ 22″), and occupies on the west coast of Scotland a position almost identical with that of the Bell Rock on the east. Nor was it less dangerous or fatal to the mariner, but yearly exacted its **tribute** of wrecked **vessels and precious** lives. A few minutes accomplished the destruction of **any** unfortunate **ship driven against its** formidable rocks, **and her** shattered **timbers were quickly borne** onward **by the ocean currents to the fishermen of the** island of **Tiree.** This remarkable survival or product of remote volcanic convulsions was not, however, **totally** submerged; **some of its** higher points rose above the **level of the** highest **tides.** But it extended its foundations over a considerable **area;** and even in the summer season it formed a constant obstruction and source of danger in the difficult **channel** between the mainland and the Outer Hebrides.

For various reasons the attention of the Commissioners of Northern **Lights** had been early directed to this formidable reef, and in 1814 they had determined to mark its locality by the erection of a lighthouse. It was visited in this same year by some of the members of the Commission, accompanied by one whose name alone is sufficient to render the visit ever memorable—Sir Walter Scott. He was much struck with the desolateness of the situation, which he thought infinitely surpassed that of the Bell Rock or the Eddystone.

Owing, perhaps, to the difficulty of the enterprise, it was deferred until the autumn of 1834, when Mr. Alan Stevenson was **authorized to commence a** preliminary inspection, **which he did not complete until 1835.** This difficulty **proceeded not only from the position but from the nature of the reef itself.**

It is true that the distance from the mainland was three miles less in the case of Skerryvore than in that of the Bell Rock; but the barren and over-populated island of Tiree did not offer the resources of the eastern coast, nor a safe and commodious port like that of Arbroath. The engineers were therefore compelled to erect, at the nearest and most favourable point of Tiree, a quay and a small harbour, with temporary cabins for the workmen, and store-houses of every kind; all whose materials, excepting only stone—and even the supply of *that* failed after a while—required to be transported from distant parts.

The first and most embarrassing, perhaps, of the numerous questions which present themselves to the engineer when entering upon the construction of a lighthouse, are those of the *height* and the *mass*. In the days of Smeaton, when the best light in use was that of common candles, the elevation beyond a certain point could not be of any utility; while in 1835 the application of the reflector and the lens, by assisting in the extension and diffusion of the light, rendered, on the contrary, a considerable elevation both necessary and desirable.

It was therefore decided that the height of the Skerryvore lighthouse should be one hundred and thirty-five feet above the highest tides, so as to command a horizon visible for a radius of eighteen miles. The diameter of the base was fixed at forty-two feet, and that of the topmost story at sixteen feet; consequently the masonry of the tower would be double that of the Bell Rock, and four and a half times that of the Eddystone.

Another peculiarity distinguishes the Skerryvore from the Bell Rock. The sandstone of the latter is wave-worn, and broken up into a thousand rugged inequalities; the action of the sea on the igneous formation on the Skerryvore has, on the contrary, communicated to it the appearance and polish of a mass of dark-coloured crystal. It is so compact and smooth that the foreman of the masons, when he landed on it, said it was like climbing up the neck of a bottle. Moreover, notwithstanding its durability, the gneiss of Skerryvore is excavated into caverns which considerably limit the area adapted for building operations. One of these caverns, we are told, terminates in a narrow spherical chamber, with an upper opening, through which, from time to time, springs a bright, luminous shaft of water, twenty feet high, and white as snow, except when the sun wreathes it with a thousand rainbows.

Mr. Alan Stevenson began actual operations in 1838 by the erection of a provisional barrack upon piles, at such a height as to be beyond the reach of all average tides. This was designed to shelter the men at night, saving them the voyage to and from the mainland, and also to accommodate them when their work was suspended by bad weather. The first erection was swept away in a great gale on the night of

November 3; but happily the labours of the season were then ended, and there were no occupants. On this occasion the grind-stone was deposited in a hole thirty-six feet deep; the iron anvil was transported thirteen yards from the place where it had been left; the iron stanchions were bent and twisted like cork-screws; and, finally, a stone weighing half a hundred-weight, lying at the bottom of an excavation, was carried to the highest surface of the rock.

Conquering all feelings of discouragement, Mr. Stevenson, in the following year, renewed his operations. A second barrack was completed by the 3rd of September. It was built of timber, and consisted of three stories: the first was appropriated as a kitchen; the second was divided into two cabins, one for the engineer and one for the master of the works; and the third belonged to the thirty workmen engaged in the erection of the lighthouse.

A more remarkable habitation than this was never dwelt in by human beings. It was an oasis in a wide waste of waters—a rude asylum suspended between sea and sky. Perched forty feet above the wave-beaten crag, Mr. Stevenson, with a goodly company of thirty men, in this singular abode, spent many a weary day and night at those times when the sea prevented a descent to the rock; anxiously looking for supplies from the shore, and earnestly longing for a change of weather favourable to the recommencement of the works. For miles around nothing could be seen but white foaming breakers, and nothing heard but howling winds and lashing waves.

In the erection of the lighthouse itself, the first important operation, and one which occupied the whole of the season of 1839—from the 6th of May to the 30th of September—was the excavation of a suitable foundation. When building the Eddystone, Mr. Smeaton had been compelled to take into consideration the peculiar structure of the rock, and to adapt his lower courses of masonry, as we have seen, to a series of gradually ascending terraces formed by the successive ledges of the rock itself. This difficult and expensive process was rendered unnecessary by the geodesical formation of the Skerryvore. Mr. Stevenson, therefore, began work by hollowing out a base of forty feet in diameter—the largest area he could obtain without any change of level. This portion of his enterprise occupied twenty men for two hundred and seventeen days; two hundred and ninety-six charges of gunpowder were made use of; and two thousand tons of *débris* and refuse were cast into the sea. The mining or blasting operations were not carried on without great difficulty, on account of the absence of any shelter for the miners, who were unable to retire more than ten or twelve paces, at the farthest, from the spot where the charge was fired. The quantities of gunpowder, therefore, were measured with the utmost nicety; a few grains too many, and the whole company of engineers and workmen would have been blown into the air. Mr. Stevenson himself generally fired the train, or it was done under his superintendence and in his presence; and from the precautions suggested by his skill and prudence, happily no accident occurred.

During the first month of their residence in the barrack, he informs us[*] that he and his men suffered much inconvenience from the inundation of their apartments. On one occasion, moreover, they were a fortnight without receiving any communication from the mainland, or from the steam-tug attached to the works; and during the greater part of this time they saw nothing but white plains of foam spreading as far as the eye could reach, and the only sounds were the whistling of the wind and the thunderous roar of the billows, which ever and anon swelled into such a tumult that it was almost impossible to hear one another speak. We may well conceive that a scene so awful, with the ruins of their first barrack lying within a few feet of them, was calculated to fill their minds with the most discouraging apprehensions. Mr. Stevenson records, in simple but graphic language, the indefinite sensations of terror with which he was aroused one night when a tremendous wave broke against the timber structure, and all the occupants of the chamber beneath him involuntarily uttered a terrible cry.

[*] A. Stevenson, "Account of the Skerryvore Lighthouse," p. 143.

They sprang from their beds in the conviction that the whole building had been precipitated to the depths of ocean.

Up to the 20th of June no materials had been landed on the rock but iron and timber. Next arrived the great stones, all ready cut and hewn, and weighing not less than eight hundred tons. But the disembarkation of these very essential supplies entailed serious risks, which were renewed with every block, for the loss of a single one would have delayed the works. At length the foundation-stone was fixed in its place; the Duke of Argyll presiding over the ceremony, accompanied by his duchess, his daughter, and a numerous retinue.

The summer of 1840 was a summer of tempests. Nevertheless, in the midst of incessant fears, and dangers, and wearying accidents, and every kind of privation, the devoted band of workers prosecuted their noble enterprise; and such, says Mr. Stevenson, was their profound sense of duty—such the desire of every one that full and complete success should crown their efforts—that not a man expressed a wish to retreat from the battle-field where he was exposed to so many enemies.

The day's occupations were thus divided. At half-past three in the morning they were awakened, and from four o'clock to eight they laboured without a pause; at eight they were allowed half-an-hour for

dinner. Work was then resumed, and continued for
seven or eight, or, if it were very urgent, even for
nine hours. Next came supper, which was eaten
leisurely and comfortably in the cool of the evening.
This prolonged toil produced a continual sleepiness, so
that those who stood still for any time invariably fell
off into a profound slumber; which, adds Mr. Stevenson, frequently happened to himself during breakfast
and dinner. Several times also he woke up, pen in
hand, with a word begun on the page of his diary.
Life, however, on the desert rock of the Skerryvore
seems not to have been without its peculiar pleasures.
The grandeur of ocean's angry outbursts—the hoarse
murmur of the waters—the shrill harsh cries of the
sea-birds which incessantly hovered round them—the
splendour of a sea polished like a mirror—the glory of
a cloudless sky—the solemn silence of azure nights,
sometimes sown thick with stars, sometimes illuminated by the full moon—were scenes of a panorama as
novel as it was wonderful, which could not fail to
awaken serious thought even in the dullest and most
indifferent minds. Consider, too—when we think of
Mr. Stevenson and his devoted company—the continual
emotions which they experienced of hope and anxiety;
the necessity, on the part of their leader, of incessant
watchfulness, and of readiness of resource to grapple
with every difficulty; the gratification with which
each man regarded the gradual growth, under his
laborious hands, of a noble and beneficent work,—and

we think the reader will admit that life upon the Skerryvore, if it had its troubles and its perils, was not without its rewards and happiness.

In July 1841 the masonry had been carried to an elevation which rendered impossible the further employment of the stationary crane. As a substitute the balance crane was introduced—that beautiful machine, invented at the Bell Rock, which rises simultaneously with the edifice it assists to raise.

Thanks to this new auxiliary, the mass of masonry completed in the season of 1841 amounted to thirty thousand cubic feet, more than double the mass of the Eddystone, and exceeding that of the Bell Rock lighthouse. Such was the delicate precision observed in the previous shaping and fitting of the stones, that after they had been regularly fixed in their respective places, the diameter of each course did not vary one-sixth of an inch from the prescribed dimensions, and the height was only one inch more than had been determined by the architect in his previous calculations.

On the 21st of July 1842, the steamer saluted with its one gun the disembarkation of the last cargo of stones intended for the lighthouse. On the 10th of August the lantern arrived, which was hauled up to its position and duly fixed, a temporary shelter from the weather being also erected for it.

The summer of 1843 was devoted to pointing the

SKERRYVORE LIGHTHOUSE.

external masonry—a wearisome operation, conducted by means of suspended scaffolds—and to the completion of the internal arrangements. And at length, on the 1st of February 1844, the welcome light of the Skerryvore pharos blazed across the waters of the stormy sea.

The light exhibited here is a white light, revolving at one-minute intervals, and distributing its radiance over an area of eighteen miles.

In 1859 the Skerryvore was visited by the members of a Royal Commission. They speak of it as the finest building they had seen, and proceed to furnish some minute particulars. The landing, they say, is by an iron ladder, and iron ways are fixed in the rock from the landing-place to the foot of the building, which have withstood the sea, and enable the keepers and persons bringing stores to move about the rock with comparative ease. The ascent to the door is by a gun-metal ladder.

The first story of the lighthouse contains water-tanks for one thousand three hundred gallons; the second, coal-bunkers for thirteen tons; the third is a workshop, used for carpentering and other avocations. The fourth story is a store-room; the fifth, a kitchen; the sixth and seventh are used for bedrooms; the eighth is a library; the ninth, an oil-store, containing one thousand and thirty-eight gallons; and the lantern forms the tenth. The bedrooms are divided into two cabins each; a lamp fixed outside gives light to each through windows. The cabins are fitted with oak, and have large looking-glasses a foot square set in panels. The library is well furnished with handsome chairs. The lantern, which is very lofty, is surrounded by a gallery with gun-metal rail; a dial is set up on the outside. " The illuminating apparatus is revolving, fixed prisms below, eight panels of lenses revolving, and eight smaller panels also revolving above, to concentrate the upper rays; these are thrown on eight plane mirrors, which deflect them to the horizon parallel to the rest of the beam. The light is, therefore, a fixed light of low power, varied by strong revolving flashes. The lamp has four wicks, and is worked by pumps which ring a small bell while in action. The lamp machinery is wound up every hour and a half, and the keepers wind the revolving machinery at the same time, though it will go for three hours. The oil is hoisted up to the top of the tower by a movable crane, the water pumped up by a force-pump. All the rooms have bell-signals, worked by blowing tubes, so that the keepers can call each other without leaving the lantern.'

A couple of fog-bells ring every minute, but cannot be heard at any great distance. The keepers here occasionally catch a few fish, such as small cod and rockfish. Occasionally a seal makes its appearance, in which case the fish beat a rapid retreat. The birds that commit suicide by dashing against the lantern panes are blackbirds, thrushes, starlings, and once a woodcock; but the number is small.

In 1866 the Commissioners of Northern Lights visited the Skerryvore. One of them was the late Dr. William Chambers, at the time Lord Provost of Edinburgh, and it occurred to him to put on record the details of this visit. On stepping ashore, they saw before them a pathway of ribbed iron riveted to the rock, and painted red, which enabled them to reach without much difficulty the foot of the tower. Here, on looking around, they perceived at least an acre of rocks in detached masses visible above the water, with a limited smooth space for walking about on all sides of the building. Dry, and free from marine plants, the higher part of the ledge is about fifteen to eighteen feet above the sea-level, and, except in very stormy weather, the rocks adjoining the lighthouse, and certain outlying patches, are never entirely covered.

Let us ascend to the interior. Climbing hand-over-hand up a weather-stained brass ladder attached to the side of the tower, we reach the doorway in the substantial wall, and find ourselves in what may be styled the ground-floor of the building. Stone is above, below, and around us, for, to prevent all risk of fire, neither ceiling nor floor shows any trace of woodwork. A step-ladder, bent to the interior curve, enables us, by clutching to a brass rail, to reach the next story above; and so on until we reach the top. In the construction of the stone floor, which is repeated in each story, there is much to admire. It consists of an arch, but not of the ordinary kind. From the walls around flat stones are projected and jointed into one central stone, the whole forming a compact mass, level on the top for the floor of one chamber, and slightly curved on the under for the ceiling of the chamber below. These flat stone arches, in which openings are left for the ladders, probably assist to strengthen the general fabric. The lower stories are used for stores of coal, fresh water, provisions, and other articles. Above are the sitting and sleeping rooms, lighted by windows, and fitted up with furnishings of oak.

Here, as elsewhere, the arrangement is for the keepers to watch four hours alternately, and on no account whatever is one to leave until another takes his place. The watcher can readily communicate with the next man on duty by blowing through a small tube in the wall, and thus setting a bell in motion. Stationed in the topmost chamber, the keeper has overhead the great blaze of light thrown out by the central lamp, which, according to the dioptric method, shines through annular lenses ; beside him, in the centre of the apartment, is the mechanism, in the form of clock-work, by which the frame of lenses revolves, and causes an alternation of darkness and a bright burst of light every minute.

"As the weather had partially cleared," says Dr. Chambers, "we had a pretty extensive view over the waste of waters from the balcony. The only visible land was that of Tiree at Hynish, with its signal-tower. I was interested in knowing the method of intercourse by signals. Every morning, between nine and ten o'clock, a ball is to be hoisted at the lighthouse to signify that all is well at the Skerryvore. Should this signal fail to be given, a ball is raised at Hynish to inquire if anything is wrong. Should no reply be made by the hoisting of the ball, the schooner, hurried from its wet-dock, is put to sea, and steers for the lighthouse. Three men are constantly on the rock, where each remains six weeks, and then has a fortnight on shore : the shift, which is made at low water of spring-tides, occurs for each in succession, and is managed without difficulty by means of the fourth or spare keeper at Hynish, who takes his regular turn of duty. According to these arrangements, the keepers of the Skerryvore are about nine months on the rock, and about three months with their families every year. But this regularity may be deranged by the weather. One of the keepers told me that last winter he was confined to the rock for thirteen weeks, in consequence of the troubled state of the sea preventing personal communication with the shore. I inquired how high the waves washed up the sides of the tower during the most severe storms, and was told that they sometimes rose as high as the first window, or about sixty feet above the level of the rocks; yet, that even in these frightful tumults of winds and waves the building never shook, and no apprehension of danger was entertained.

"When the weather is fine, the keepers are not by

any means confined to the building. They may straggle about among the gullies, enjoy the fresh air, and amuse themselves by angling for the smaller kinds of white-fish, any catch of this sort imparting a little relish to the monotony of the daily fare. The visits of seals, which are occasionally seen frisking in the surf, also furnish some amusement, and one can fancy that, to a student of natural history, life at the Skerryvore might furnish some useful memoranda. The keepers do not complain of solitude; the obligations of professional duty, and the periodical return to their families at Hynish, where in fine weather they occupy themselves with their gardens, help materially to banish the sense of loneliness. Besides, as we observed from the visitors' book, yachting parties sometimes land on the rock and ascend to the top of the lighthouse, perhaps leaving behind them the acceptable gift of a few newspapers, to show what is going on in the outer world."

The total cost of construction of the Skerryvore lighthouse was £86,977, 17s. 7d.

The next lighthouse which we encounter, that on the *Dhu-heartach Rocks*, has also an interesting history.

The Dhu-heartach, Dubh Artach, or St. John's Rock, is situated in lat. 56° 8', about midway between the Skerryvore and the Rhinns of Islay—that is, about twenty miles from Islay, eighteen miles from Colonsay, fifteen miles from Iona, and fifteen miles from Mull—

in the centre of an archipelago which ancient legend and ecclesiastical history and modern romance have combined to invest with precious associations. Geologically speaking, it is an irregular mass of the dark-green basaltic or pyroxene rock known as augite; and it measures two hundred and forty feet in length by forty-three feet in breadth, with a rounded summit rising forty-seven feet above high-water mark. Its surface is deeply furrowed in all directions, and is almost entirely bare of vegetation. There is neither cove nor creek in which a boat can be moored, and its black walls on every side start up sheer from the sea. These walls are solid with an almost impenetrable solidity; and need be so, for the Atlantic spends its full and unchecked violence upon them. It is difficult of approach even in fair weather; while in times of storm it is inaccessible, the ocean sweeping over it with great tracts of foam, with a fury intensified by the conflicting currents which pour through the island-channels of the Hebrides.

During the severe gales of the winter of 1865–6 many ships were lost in this dangerous neighbourhood, where the sea is literally honeycombed with sunken rocks and hidden reefs; and the Commissioners of the Northern Lighthouses, with the sanction of the Board of Trade and the Trinity House, resolved to plant a warning and a guiding light on the Dhu-heartach. The work was intrusted to Messrs. D. and T. Stevenson of Edinburgh, who estimated the cost at £56,900. In form it is a "parabolic frustum," the topmost course of which rises one hundred and nine feet above the base. The diameter diminishes, as it ascends, from thirty-six feet to sixteen feet. There are seven stories besides the light-room. The total height of the lantern above the sea is one hundred and forty-five feet, commanding a range of about seventeen miles. It shows a fixed white light, with a red sector. In foggy weather a bell is rung for ten seconds at intervals of half a minute.

We gather some noteworthy particulars from Mr. Scott Dalgleish's graphic account in *The Times* of a visit he paid to the Dhu-heartach in September 1881. "If it be true," he says, "that lighthouse-building involves heroism and self-sacrifice, it is no less true that lighthouse-keeping requires the exercise of the same rare qualities in a degree hardly less striking. Not only our lighthouse men, but also those whose duty it is to supply them with the means of subsistence and of discharging their important duties, at all occasions and in all weathers, deserve well of the country, especially of a maritime nation like Great Britain, which depends so largely for its livelihood on those who go down to the sea in ships......As we approached the lighthouse, we could see the keepers at work on the lee-side getting ready the derrick by which we were to be landed. Up to this time it had been doubtful—so said both the captain and the mate—whether the sea would be calm enough; but when we got under the lee of the island, we were assured that landing was quite practicable, though to our unaccustomed eyes the lumpy sea breaking in white waves on the rock did not promise much comfort. **When we** were within **a** furlong of the island the little steamer dropped anchor. The longboat was launched, manned by four sailors, and steered by Captain Irving himself, and in this we were rowed **to** the south-eastern end of the rock. The process of landing is interesting, though when experienced for the first time it must appear rather sensational to those **of weak nerves.** The boat is not allowed to touch the **rock.** It is anchored **by a long line** stretching seawards, so as to allow its stern **to** swing within ten or twelve **feet** of the rock. The boat is kept in position under the **derrick by two stern** lines attached to the rock. The derrick consists simply of a spar, which is rigged up by the lighthouse-keepers as often as it is required, and from which **a stout** rope working **in a** double pulley is suspended. When the boat is in position, the rope, which **has a** loop at the end of it, is dropped into the stern. **You put one foot** into the loop, hold tightly to **the rope with** both hands below the block, and are first hoisted into the air and then pulled downwards to the rock. There you are clasped in the strong arms of one of the keepers; and before **you** are released from the friendly grip, you are reassured by **a kindly voice** bidding **you** 'Welcome to Dhu-heartach!'

"The view of the lighthouse **from the rock is totally**

different from that obtained of it from the sea at a distance of a mile. In the latter view you take in at once all its proportions; and while the frustum which forms its base is, perhaps, too broad to give the notion of elegance, the upper part stands out from the sky as a slender and graceful shaft. When you see the lighthouse from the rock on which it stands you lose the general outline; you see only the massive details, and the one idea impressed on the mind is that of tremendous strength. This idea is intensified when you walk round the granite cone, which seems as immovable as the rock in which it is securely embedded. The diameter at the base is thirty-six feet, and for the first thirty courses the cone consists of blocks of granite dovetailed and 'joggled' into a solid mass. The topmost course of masonry, where the diameter is only sixteen feet, is one hundred and nine feet above the base, and on that rests the cupola or lantern in which the light is enclosed......Around the lantern there is an open gallery, formed by a strong iron railing, from which magnificent views are obtained of the sea and the distant islands. Below the light-room there are six apartments, each occupying a story. The lowest story is about forty feet above the rock, and access to it is obtained by an outside ladder of gun-metal steps fixed in the granite wall. The ascent of this ladder is even more trying to weak nerves than the process of landing, or being landed, on the rock with the derrick. Safety lies in keeping a firm hold with hands and feet,

and looking upwards rather than downwards. When
you have made the ascent, you are in the coal and
water store. You look around for a minute or two to
realize your situation, and then you mount by a steep
and narrow flight of oaken steps to the oil-store.
There is nothing to detain you there; so you ascend
by another wooden staircase to the kitchen—a trim
little room, with a neat close range on which a copper
kettle simmers, several cupboards, a table and chairs,
a carpenter's bench hung from one side, and two plate-
glass windows set deeply in the granite wall. Above
that, and reached in the same way, is the dry-store, in
which the provisions are kept. Above that, and still
smaller—for, of course, the area contracts as we ascend
—is the sleeping-room, in which there are six berths,
arranged in two tiers of three in each. Above is the
library—a pleasant little sitting-room, furnished with
cupboards filled with books and periodicals, and with
cheerful outlooks seawards. Above the library is the
lantern or light-room, which is in some respects the
most interesting apartment in the building. From the
topmost course of granite masonry, which is here two
feet thick, rises a circular glass-house, consisting of
large diamond-shaped lunettes of plate-glass fixed in
metal frames. Within this is the lantern proper, which
is a fixed, circular white light, consisting of convex
lenses and prisms arranged on the dioptric principle.
The light appears as a white light all round, except on
the south-east, where there is a red sector, produced,

not by any change in the lantern itself, but by the
outer wall of the cupola being filled in with red glass.
At present, the red light, as well as the white light, is
stationary; but it has been resolved to make the red a
flashing light......Between the paraffin lamp and the
dioptric lenses which surround it five men could stand
with ease. Connected with the machinery which supplies the lamp with oil there is a small bell, which
rings every second 'while the lamp holds on to burn,'
and stops as soon as the supply of oil fails. This con-
trivance is designed to prevent the attendant watcher
from falling asleep. At first the bell was made to ring
when the lamp went out; but it was thought that, if
the lamp went out, the tinkling of a small bell in the
midst of the storm might fail to awaken him, and, on the
principle that prevention is better than cure, the present arrangement was devised. Even if the watch fell
asleep while the bell was ringing, its stopping would
in all probability awaken him. There is not much
likelihood, however, of the arrangement being subjected to this test. There are always three men in the
lighthouse, and as during the night each of them
watches for three hours in the lamp-room, each of
them may also rest for six. One of the most important and laborious of the watcher's duties, at least in
foggy weather, is to attend to the machinery connected
with the fog-bell—a large and powerful bell suspended
in the gallery outside the lighthouse on the level of the
lamp-room. Its purpose is to appeal to seafaring wan-

derers **through the** sense **of** hearing when the light is hidden from their view. **The** bell rings for ten seconds at a time, with intervals **of** thirty seconds. **The ma-** chinery **on** which its action depends requires to be wound up every twenty minutes. This makes constant **demands on** the attention, **and also on the** physical **powers, of** the **attendant, for the weight to be** raised **is heavy, and** there is hardly sufficient room in the **lantern to** turn the winch."

We now arrive at *Ardnamurchan Point*—that is, "the Point of the Great Seas"—a bold promontory in Argyllshire, and the westernmost headland on the Scottish coast, **forming** the northern **boundary of the mouth of Loch** Sunart. **The** tall white tower **of** granite, erected here in 1849 **by** Mr. **Alan** Stevenson, **measures** one hundred and eighteen **feet from base to** summit, and carries **a fixed** white **light (one hundred and** eighty feet above **the sea), which extends its** protection **over** six leagues **of rolling waters. Off** this **point** the sea is **always heavy. Cost £13,738, 0s. 10d.**

The Sound of Mull has its guiding light **on the** *Runa Gall Rock,* **in lat. 56° 38′**—a mass of columnar basalt containing agates. The lighthouse, sixty-three feet high, built **of** brick and whitened with stone mouldings, dates from **1857, and** was designed by Messrs. Stevenson. **It** carries a dioptric holophotal **light,** fixed, **red, to seaward, to** distinguish it from the

neighbouring light of Ardnamurchan, green towards
the opposite shore, and white towards the Sound of
Mull. The keepers here are well lodged; and the
tower is furnished with the usual instruments—clock,
dial, telescope, barometer, thermometer, rain - gauge,
lightning-conductor. Cost £6,277, 15s. 7d.

At *Mousedale Island*, off the coast of Lismore, in
lat. 56° 27′ 19″, we find a light-tower of stone, eighty-
six feet high, erected by Mr. R. Stevenson in 1833,
which exhibits a fixed white light, visible at a distance
of sixteen miles. Cost £11,299, 10s. 5d.

Thence we stretch across to *Corran Point*, in lat.
56° 43′ 16″, where Loch Eil branches from Loch
Linnhe, and find there a lighthouse on a smaller
scale, erected in 1860. Its height is forty-two feet.
It carries a fixed red light, with a white sector,
which can be sighted by the mariner within eleven
miles.

Fladda Island lies off the Argyllshire coast, between Oban and the mouth of the Crinan Canal, at
the north end of Suerba Sound. The lighthouse here
aids the navigation of the channel of the Dorisht-mhor,
or Great Gate, between Craignish Point and the mainland. It is forty-two feet in height, and was erected
in 1860. Shows a fixed white light, with a red sector,
and can be seen eleven miles off.

In the Sound of Jura lies the *Skeir Maoile* or *Iron Rock* (lat. 55° 52′ 30″)—almost in the same parallel as the St. Abb's Head Lighthouse, on the east coast—where the *Chevalier* steamer was lost about thirty years ago. It had long been a dangerous obstacle to the navigation of these waters; but the erection of a lighthouse upon it, though much desired by ship-owners and seamen, was delayed until 1865, when the present handsome edifice, eighty-three feet in height, was built at a cost of about £10,000. It shows a white light, revolving every thirty seconds, and visible fourteen miles.

We next arrive at Islay Island, where a lighthouse was erected at *Rhu Vaal* or *Rudha Mhail* in 1859. It is an exceedingly handsome structure of brick, was designed by Mr. Stevenson, and measures one hundred and thirteen feet in height from base to vane. Its light, white and fixed, with red sector, shows over the outer end of Oronsay for a considerable distance from the point. It opens up the northern entrance to the Sound of Islay, and is also serviceable for the navigation of the channel between Oronsay and Colonsay. Cost £7,437, 4s. 9d.

But for the better lighting of an admittedly dangerous stretch of sea, a lighthouse, two years later, was set up at the south end of Islay Sound, on the summit of *Macarthur Head*, lat. 55° 45′ 50″. The building is forty-two feet high—the centre of the lantern one

hundred and twenty-eight feet above high-water mark
—and it exhibits a fixed white light, with a red sector,
the fair-weather range of which is eighteen miles.

In 1825 a first-class lighthouse, ninety-six feet in
elevation, and built of stone, was raised by Mr. Robert
Stevenson, on *Oversay Island*, in the Rhinns of Islay,
lat. 55° 40′ 20″. The lantern has twenty-five reflectors,
revolving, showing a flash from these reflectors every
five seconds, which travels over eighteen miles. The
oil burned is colza, and it is stored in a cellar underground, being admitted into the interior through a
tube, with every precaution for safety. Cost £8,056,
6s. 5d.

On *Dun Point*, in Loch-in-Dail, is shown a fixed light,
with white and red sectors; and there is a square tower,
sixty feet high, dating from 1853, at *Carraig Pladda*
point, on the west of the entrance to Port Ellen. It
shows a fixed red light. The *Mull of Cantyre*, the
south-west headland of the peninsula of Cantyre, has
been lighted since 1787. The stone tower, designed by
Smith, is thirty-eight feet high, and the lantern is two
hundred and ninety-seven feet above high water. The
light, a fixed white light, is visible twenty-four miles.
This is a fog-siren station.

On the *Ship Rock of Sanda*, in lat. 55° 16′ 30″,
a stone lighthouse, forty-eight feet high, was erected
in 1850 by Mr. Alan Stevenson, at a cost of £11,931,
10s. 2d. It shows a white occulting light, visible

for eight seconds, and then dark for sixteen, with a fair-weather range of eighteen miles. A fog-siren is in use here.

We are now making for the mouth of the Clyde, and are guided onward by the Davar and Pladda lights. On *Davar Island,* in lat. 55° 25′ 45″, a tower of granite,

SHIP ROCK OF SANDA LIGHTHOUSE.

with an elevation of sixty-five feet, was erected in 1854. It exhibits a white light, revolving at half-minute intervals, and visible for seventeen miles. Two lights are shown on the little islet of *Pladda,* off the south-east point of Arran. One of the lighthouses is no less than ninety-five feet high; the other, forty-three feet high; both first lighted in 1790. They were designed by Mr.

Thomas Smith, and built of stone. Both lights are white and fixed; the higher can be seen at seventeen miles, and the lower at fourteen miles, according to the direction in which ships are approaching. A fog-horn is sounded here in thick weather.

The Clyde is, as a matter of course, well lighted, and has besides a very full and skilful system of beacons and buoys. We shall run up it only as far as the *Cloch*—a lighthouse with which river-excursionists and West Highland tourists are perfectly familiar. This shapely white tower (of freestone), eighty feet high, and elevated seventy-six feet above high water, stands upon Cloch or Clough Point, just opposite Dunoon, in lat. 55° 56′ 35″, and was erected in 1797. Its fixed white light is visible seven miles down the river. A couple of steam-whistles of different pitch are used as a fog-signal, giving a blast every seven and a half seconds.

There is also a lighthouse on *Toward Point,* where the Clyde runs into the picturesque winding channel of the Kyles of Bute. Built in 1812 from Mr. R. Stevenson's designs; is of freestone, with inner and outer walls, and has keepers' dwellings adjacent; is close to the village of Innellan. From its position, its inmates are not exposed to any of the perils which environ such stations as the Eddystone or the Bell Rock, though they sometimes get a pretty stiff gale about their ears. The tower is sixty-three feet high, with a brilliant white light, flashing every ten seconds, and visible fourteen miles.

Nearer the river-mouth, and between the island of

Bute and the Ayrshire coast (below Largs), lie the rocky masses of the two Cumbrae Islands—*Great* and *Little Cumbrae.* There is a lighthouse on the west coast of the latter. A coal-fire in a chauffer was maintained on the highest point of this island contemporaneously with that on the Isle of May, and in 1750 was replaced by a lighthouse, a circular tower about thirty feet high. From its situation it was, however, so frequently hidden in fog and mist as to be of imperfect utility to mariners ; and in 1757 a new building was erected (repaired in 1829 and 1836) in its present position on the west side of the little island. This is only thirty-nine feet in height, and ungraced by any architectural beauty. The lantern, which rises one hundred and fifteen feet above high-water mark, is equipped with a fixed white light, visible sixteen miles in clear weather. A fog-trumpet emits blasts of five seconds' duration, with intervals of eighteen to twenty seconds between each.

Leaving the Clyde estuary, and continuing our coast survey, we steam past the lights of Ardrossan, Irvine, Troon, and Ayr harbours, and arrive off *Turnberry Point,* a headland on the bold and rocky coast of Carrick. You can see the gray, mouldering ruins of Turnberry Castle, the seat of the powerful Lords of Carrick during the twelfth and thirteenth centuries, the famous old fortalice

" Where Bruce once ruled the martial ranks,
And shook his Carrick spear."

A lighthouse of white brick, sixty-four feet in height, was erected here in 1873, which exhibits a white light, flashing every twelve seconds, and visible for fifteen miles.

From this point we obtain a fine view of the conical mass of *Ailsa Craig*, which lies off the mouth of the Firth of Clyde, between the coasts of Ayrshire and Cantyre, in lat. 55° 15′ 10″ and long. 5° 6′ 15″. This island-mountain of columnar syenitic trap shoots up to an elevation of four hundred feet, from a base of three thousand three hundred feet by two thousand two hundred feet. Its formation is distinctly columnar, especially on the western side, where it rises sheer from the sea. It is inhabited by a few rabbits and goats, and by thousands of solan-geese, puffins, cormorants, auks, and gulls. On the east side of this " craggy ocean pyramid," and on a low projecting spit, a lighthouse tower, thirty-six feet high, and about sixty feet above high-water mark, was erected in 1886. It shows a white light, flashing in quick succession for fifteen seconds, and then eclipsed for as many. At the south end of the island a fog-siren is sounded in foggy weather, giving three blasts in rapid succession every three minutes; while at the north end another siren gives a five-seconds blast every three minutes. These two signals are so arranged as to begin to sound about one minute and a half after each other.

On *Cairnryan Point*, at the mouth of Loch Ryan —an arm of the sea leading up to Stranraer—a stone

lighthouse, fifty feet high, was erected in 1847 by Mr.
Alan Stevenson, at a cost of £4,241, 15s. 5d., with a fixed
white light, visible twelve miles. But on the other
side of the loch, whence projects the low headland of
Corsewall *Point*, Wigtownshire, a more imposing building was raised in 1817, from the designs of Mr. Robert
Stevenson. Its height, from base to vane, is one hundred and ten feet; it is of stone, and it exhibits a
revolving light, white and red alternately, every minute,
with a fair-weather range of sixteen miles. Cost £7,835,
19s. 8d.

The extreme southern point of Scotland is the picturesque rocky promontory of the *Mull of Galloway*,
in lat. 54° 38′ 10″—almost in a parallel line with
Whitby, on the north-east coast of England. About a
mile and a half long, and a quarter of a mile wide, it is
connected with the mainland by a long narrow isthmus,
the sides of which curve into two small bays, called
respectively East and West Tarbet. On the south and
south-west the rocky walls of this bold headland spring
from the sea almost perpendicularly to an elevation of
two hundred and three hundred feet, and are hollowed
by caverns, in which the foaming waters, with a
southerly wind and a flowing tide, roll and roar like
reverberations of thunder. Here was erected, in 1830,
from Mr. Robert Stevenson's plans, a massive lighthouse of stone, eighty-six feet high, with double walls,
which now displays a white intermittent light (visible,
that is, for thirty, and then eclipsed for fifteen seconds)

over the extended range of twenty-five miles. It is distant twenty-one miles north-north-west from the Point-of-Ayre lighthouse, in the Isle of Man, and about the same distance, south-east by east, from Copeland Lighthouse, on the coast of Ireland. The balcony of the tower commands a magnificent and most extensive view of the lofty summits of the Southern Highlands of Scotland, the towering Paps of Jura, the wide and shining expanse of the Irish Sea, some thirty leagues of the green coast of Erin, the bold outlines of the Isle of Man, and the remote peaks of Cumberland—a picture not easily surpassed in brilliancy of colouring and variety of interest. The cost of this building was £8,378, 9s. 9d.

The last lighthouse on the coast of Scotland is on *Little Ross Island,* at the mouth of the Solway, in lat. 54° 46′. The building is a substantial tower of stone, with double walls, erected in 1843, from Mr. Robert Stevenson's designs; of a total height of sixty-five feet, and furnished with a white light, flashing every five seconds. Cost £8,478, 15s. 7d.

THE ISLE OF MAN.

As the lights which protect the coasts of the Isle of Man are under the control of the Commissioners of the Northern Lighthouses, it will be convenient to notice them here. They are four in number—Ayre Point, Chicken Rock, Langness, and Douglas Head.

The *Ayre Point* or Point-of-Ayre lighthouse, in lat.

54° 24′ 56″, is a stone tower, ninety-nine feet high, erected in 1818 from Mr. R. Stevenson's designs. It is situated in the middle of a sandy common, about a quarter of a mile from the sea, and displays a revolving light, red and white alternately, at one-minute intervals, with a fair-weather range of sixteen miles. The keeper at this station can see the lights at the Mull of Galloway, Little Ross Island, St. Bees, and the *Bahama Bank* lightship.

The *Chicken Rock* lies in lat. 54° 24′, about three-quarters of a mile south of the Calf of Man, and is exposed to the tremendous force of the Atlantic billows when they are driven through St. George's Channel by south-westerly gales. At high water very little of its surface is visible; but at low water the receding tide leaves exposed a rugged tract of nearly eight thousand feet square, and it is then seen to be formed by a couple of rocky islets, north and south, united by a low isthmus. In the midst of a rapid tideway, and in comparatively deep water, the Chicken Rock was long an object of dread to seamen, more particularly as, in foggy weather, the lights then shown on the Calf of Man* were seldom visible. In 1869, therefore, the Northern Commissioners, with the sanction of the Board of Trade, removed the Calf of Man light, which was practically useless, and ordered the erection of a first-

---

* Two lighthouses of stone, seventy feet and fifty-five feet high, designed by R. Stevenson, and built in 1818. Cost, including the Point-of-Ayre lighthouse, £18,846, 16s. 3d.

class lighthouse on the Chicken Rock. The tower is
of granite, and strong and graceful, as all Messrs.
Stevenson's lighthouses are. It was completed in
December 1874, and on the first of January 1875 the
welcome ray of guidance and warning shone from this
"Tadmor of the wave," to the great advantage of navigation.

The labour of preparing the stone was carried on at
Port St. Mary's, a little fishing-village about four miles
and a half from the rock. There, too, the workmen
took up their quarters, a steam-tug being employed to
convey them to and from the rock. In order to land
them she carried two large quarter-boats; while, as a
precaution against possible accidents, every man was
ordered to wear a cork life-belt on every trip, no matter
what might be the condition of the sea. The work
was necessarily regulated by the tides and the weather:
sometimes it was continued for six or seven hours;
sometimes it was suspended at the end of one or two.
Nothing, of course, was done during the winter—that
is, from the end of September to the beginning of April.
About thirty-five men were generally employed. Nine
of the lowest courses of masonry were laid in 1870;
fourteen more in 1871. Thus far—that is, to an elevation of thirty-two feet eight inches above the rock—the
tower was solid. Forty-seven courses were added in
1872. The ninety-sixth and topmost course was laid
on June 6th, 1873; and in the following year the
internal fittings were completed. The lighthouse pre-

sents the form of a noble and shapely column of gray granite, one hundred and forty-three feet in height from base to vane. Its outline is that of a "hyperbolic curve," which is recommended by many practical advantages. The interior consists of eight stories: the lowest, store-room for coal and water; second, store-room for oil; third, dry stores; fourth, kitchen; fifth, provision-room; sixth and seventh, bedrooms; eighth, light-room. The sashes of the lantern are ten feet high, and fitted with the best plate-glass—thick enough to resist the heedless wing of any sea-bird attracted by the illuminated pane. The lighting apparatus is Fresnel's, with Mr. Stevenson's holophotal improvements. An octagonal metal frame, each side of which is provided with a large annular lens, revolves round the central lamp every four minutes, with the effect of producing a steadfast intense glow or beam of white light once every thirty seconds, as each annular lens passes in front of the lamp. In fair weather the range of illumination extends to eighteen miles. Here, as at other first-class lighthouses, two bells suspended from the balcony are rung during fogs at intervals of half a minute.

The total cost of the Chicken Rock lighthouse was £64,559.

On the south-eastern side of *Langness* a lighthouse was erected in 1880; a circular limestone tower, sixty-three feet high, which displays a white light, flashing, for five seconds. This is a fog-siren station.

The stone tower on *Douglas Head,* lat. 54° 8′ 35″, dates from 1832. It is sixty-five feet in height from base to vane, with a fixed white light, visible fourteen miles; is built of slate rock; was designed by Mr. Thomas Brine; and cost about £2,500.

The *Bahama Bank* lightship, lat. 54° 19′ 40″, shows two flashes (white) in quick succession every thirty minutes, and carries a fog-siren.

ENGLAND: WEST COAST.

On the low, sandy headland of *Skinburness* or *Cote,* to the north-east of Silloth, a wooden lighthouse, thirty-two feet in height, was erected in 1841, with a fixed white light, visible for nine miles. The *Lee Scar Rock* lighthouse, south-west of Silloth, in lat. 54° 52′, is constructed upon piles, with the top of its lantern about forty-five feet above low-water mark. In Robin Rigg Channel the *Solway* lightship shows a fixed light, visible only for six miles. Passing the lights of Maryport, Workington, Harrington, and Whitehaven, we come to the lofty headland of *St. Bees,* in lat. 54° 30′ 50″—nearly the same as that of Spurn Head, on the east coast—where in 1718 was exhibited a coal-fire, and in 1872 was built a circular white lighthouse tower of sandstone, fifty-five feet high, and measuring three hundred and thirty-six feet in elevation from high-water mark to centre of lantern. Designed by Mr. J. Nelson. It has an occulting white light, which is shown for twenty-four seconds, and eclipsed for two seconds; then shown for two, and eclipsed for two. Visible twenty-five miles.

On the *Selker Rocks*, three miles north-west of Selker Point, lat. 54° 16′ 5″, a lightship was moored in 1883, which shows one white and one red flash in swift succession every half minute.

Several lights warn the seaman from the dangerous sands of Morecambe Bay, and open up the channels into the Mersey and the Dee.

The *Wyre River* lighthouse, built upon screw-piles in 1840, is situated at the north-east elbow of North Wharf Bank, where it shows a fixed white light, with a ten miles' range. A remarkable, and probably an unprecedented accident occurred here on the morning of Saturday, February 19th, 1870. "About half-past ten the schooner *Elizabeth and Jane*, of Preston, approached the mouth of the channel opposite Fleetwood. Adjoining the channel-mouth, and about three miles from the latter town, is situated the lighthouse, upon screw-piles. When about half a mile off the lighthouse, the captain of the schooner found he was drifting towards it; and, spite of all his exertions, he was unable to change her course, as the tide flowed rapidly inwards, and a dead calm prevailed. Before the anchor could get a hold, the ship was bow foremost into the piles, which were all shattered by the collision, and taking up the body of the lighthouse—a huge rectangular wooden frame, filled in with windows, and surmounted with a large revolving [fixed] light—carried it away on her forecastle. Two keepers were in the lighthouse, but neither was hurt. The vessel, however, was greatly injured, and some alarm was felt lest she should sink. However, the accident was seen from the shore; a tug-steamer came to her assistance, and, with the lighthouse on board, she was towed into port." The lighthouse was rebuilt with as little delay as possible.

On *Crosby Point,* a square tower of brick, seventy-four feet high, was erected in 1856 by Mr. Jesse Hartley. It exhibits a fixed white light, with a range of twelve miles. On the shore at *Leasowe,* midway between the Mersey and the Dee, a circular, tapered tower, one hundred and ten feet from base to vane, dates from 1763; and a stone tower, sixty-eight feet, erected in 1771, repaired in 1873, one hundred and sixty feet above high water, crowns *Bidston Hill,* in lat. 53° 24'. Both these are fixed white lights, visible at fifteen and twenty-three miles respectively. A light-house, ninety-four feet high, at *Rock Point,* restored in 1877, shows a revolving white light every half minute. There are also the *North-West,* the *Bar,* the *Formby,* the *Crosby,* and the *Dee River* lightships; all with fixed white lights, except the Formby, which displays a revolving red light.

We pass the lights of Rhyl, and at Llandudno observe a square castellated lighthouse of gray stone, on the north spur of the grandly massive promontory of the *Great Orme Head.* It was erected in 1862, and is elevated three hundred and twenty-five feet above high-water mark. Exhibits a fixed light with white and red sectors, which has a range of twenty-four miles in clear weather.

The Menai Strait, between the Welsh mainland and Anglesey, is about twelve miles long and from half a mile to three-quarters of a mile broad. To assist mariners, a lighthouse was erected in June 1838 on *Trwyn-Du* or *Black Point*, lat. 53° 18' 51"—a circular castellated tower of stone, solid to about thirty feet from foundation, and ninety-six feet in height, from which a red light is shown.

At *Lynus Point*, on the south coast of Anglesey, a castellated lighthouse, thirty-six feet high, displays a white occulting light. It was erected in 1834–6, the want of a light having been proved by the disastrous wreck of the *Rothesay Castle* on Puffin Island in 1831.

Off Holyhead, in lat. 53° 25' 15", lies *Skerries Island*, a low flat tract of land covered with sea-birds, chiefly terns. The lighthouse is a circular tower, of Anglesey stone, seventy-five feet high, dating as far back as 1714. In 1804 it was raised twenty-two feet, and oil-light shown.* Fogs here are very troublesome, lasting sometimes for forty-eight hours at a stretch. The lights are two—intermittent white, and a fixed red light, shown from a window fifty feet below the lan-

---

* When purchased by the Trinity House in 1841, the enormous sum of £444,984, 11s. 3d. was paid to the proprietor.

tern. A fog-siren is sounded when required. In 1855 the ship *Regulus* was wrecked on the Skerries.

HOLYHEAD LIGHTHOUSE.

Off the north-west point of Holyhead Island, in lat. 53° 18′ 30″, rises the *South Stack*, an isolated rock, lying immediately below the cliff, and now connected with the mainland by a bridge. In 1809 a lighthouse was erected here—a circular tower of stone, eighty-four feet high, designed by Mr. David Alexander,* which

* Lighthouse, lantern, and apparatus cost £11,828, 17s. 9d. ; bridge, £1,046, 11s. 8d. ; dwellings, £1,509.

exhibits a white light, revolving at one-minute intervals, and visible for twenty miles. The focal plane is two hundred and one feet above the sea. During foggy weather, a large fog-bell, inverted, is rung by machinery; and a smaller revolving light is occasionally shown about forty feet above the sea, at a point thirty yards north of the main lighthouse. A gun or explosive charge is also fired every five minutes.

Here, and at the Skerries, the sea-birds are preserved as a kind of natural fog-signal; they are quite tame. Gulls sit on the walls, and close to the lighthouse, screaming continually; and the visitor will be amused to see some white rabbits sitting among the young gulls, apparently on terms of perfect intimacy.

Lightships are stationed in Carnarvon and Cardigan Bays. There is a lighthouse, a massive square-built tower of gray marble, erected in 1821, from Mr. J. Nelson's designs, and newly lighted in 1885, on *Bardsey Island*, lat. 52° 45′. Height, one hundred and two feet. Shows a white light for twenty-seven seconds; eclipsed for three. A fog-siren is stationed here. Cost, with buildings, £5,470, 12s. 6d; lantern, £2,950, 16s. 7d. In 1877 a lighthouse, a circular white building thirty-five feet high, was set up on St. Tudwall Island.

In lat. 51° 51′ 10″, below New Quay, and close to Milford Haven, a circular light-tower of stone, thirty-six feet in height, designed by Mr. Walker, crowns the summit of the *South Bishop Rock*, which rises a hundred feet above high-water mark. It is furnished with a white light, revolving every **twenty** seconds, **and having a fair-weather** range **of eighteen miles. The sea off this point is** frequently very boisterous, and the spray occasionally strikes **the** lantern, notwithstanding its elevation, and has been **known to break the lower windows of the dwelling-house. The lighthouse and buildings cost £11,255, 5s. 11d. ; the** lantern, etc., **£1,493, 8s. 6d.**

The *Smalls* **are a cluster of rocks at** the **entrance to St. George's Channel, in lat. 51** 43′ 20″. **Here, in 1778, a lighthouse was first** erected **by Mr. John** Phillipps, **a Liverpool Quaker, as "a great** and **holy good to** serve **and save** humanity." But, in this instance, benevolence met with its **due** reward soon **after-wards ; and the toll** derived **from** passing vessels proved so profitable **that, when the light was surrendered to** the Trinity **House in 1836, Mr.** Phillipps's **descendants** received **no less a sum than** £170,468 as compensation.

The task which **Mr. Phillipps** had **with so much philanthropy** taken upon himself **was attended with considerable** difficulties. **In ordinary** weather **the** Smalls **Rock rises twelve feet above the** water ; **but when the seas run heavy, it is completely** submerged, **and the erection of a lighthouse** must, **in such circum-stances,** demand **the exercise of great** skill **and judg-ment.** Mr. Phillipps **was for some** time **engaged in looking for a man** capable **of** carrying **out his idea; but at** length **he** settled **upon a** Liverpool **musical**

instrument maker, named Henry Whiteside, who, in the summer of **1772**, began operations at the head of a gallant little band of Cornish miners. They had scarcely laid the foundation before the weather suddenly grew tempestuous, and the tender which waited upon them was compelled to weigh anchor and run out to sea. The unfortunate workmen clung to the rock with all the energy of despair, and remained in this wretched position for two days and nights. Whiteside, however, was not discouraged by this rough experience, nor by any of the misadventures which from time to time befell him, and in 1778 succeeded in completing his work —a timber-built lighthouse, supported upon oaken piles, forty feet high, with a total height of seventy-one feet.

One day, the fisher-folk on the mainland, which is about twenty miles distant, picked up on the beach a small barrel, inscribed with the characters, rudely painted :—

"Open this, and you will find a letter."

They obeyed the injunction, and inside the cask discovered a carefully sealed bottle, and in the bottle the following letter :—

"THE SMALLS, *February 1st, 1777.*

"SIR,—Being now in a most dangerous and distressed condition upon the Smalls, do hereby trust Providence will bring to your hand this, which prayeth for your immediate assistance to fetch us off the Smalls before the next spring [tide], or we fear we shall perish ; our water near all gone, our fire quite gone, and our house in a most melancholy manner. I doubt not but you will fetch us from here as fast as possible ; we can be got off at some part of the tide almost any weather. I need say no more, but remain your distressed

"Humble servant,

"**H. WHITESIDE.**"

Beneath this signature a postscript had been sub-joined :—

" We were distressed in a gale of wind upon the 13th of January, since which have not been able to keep any light ; but we could not have kept any light above sixteen nights longer for want of oil and candles, which makes us murmur and think we are forgotten.

" EDWARD EDWARDS, G. ADAMS, J. PRICE.

" *P.S.*—We doubt not that whoever takes up this will be so merciful as to cause it to be sent to Thomas Williams, Esq., Trelithin, near St. David's, Wales."

There is, however, a sadder page than this in the brief history of the Smalls Rock Lighthouse, and one which bears a close resemblance to an episode in the history of the Eddystone. Early in the present century, when the winter proved to be of exceptional severity, the keepers were cut off from all communication with the land for a period of four months. It was in vain that ships were sent out to the rocks ; a raging sea invariably prevented their approach. At last one of them returned with the startling intelligence that her crew had observed a man standing, upright and motionless, in a corner of the outer gallery, with a flag of distress floating beside him. Every night thenceforward the folk on the shore watched eagerly to see if the light were still kindled, and every night the welcome ray rose on the horizon punctually—a proof that there was still a keeper at the Smalls. But were *both* the keepers living ? It is needless to say that public feeling was more deeply stirred as day after day passed without any news from the lighthouse being received.

At last, in an interval of calm, a Milford fisherman succeeded in landing on the rock, and in carrying back to Solva the two keepers. But one of them was a corpse. The survivor had made a kind of shroud for his dead comrade, and afterwards placed the body in the gallery, fastening it there securely. This device he adopted in order to prevent the effluvium which would otherwise have made the light-tower uninhabitable, and yet to preserve the corpse for medical examination, lest any suspicion of foul play should be entertained.

In 1859, the Trinity House began the erection of a new and exceedingly handsome structure of stone, which, while it was in course of building, was visited by the Royal Commissioners. During their inspection, in 1859, they took note of some interesting particulars, which may be quoted here in illustration of the vast progress that within the last century has taken place in lighthouse architecture.

But first, for some details as to a lighthouse keeper's life. The head-keeper had been eighteen years, he said, on this station, and preferred it to any other. He was a Welshman by birth, was married, and had a considerable farm on shore. The under-keeper was a native of Ealing, a watchmaker to trade, and "would rather be anywhere on shore at half the money." He said, "This is rusting a fellow's life away." And, certainly, it is not easy to imagine a greater contrast to a watchmaker's life at Ealing than a lighthouse keeper's life on the solitary rock of the Smalls, twenty miles from land! The head-keeper seemed to have found diversion in bird-keeping; he said that he had caught woodcocks in September, as also larks, starlings, and blackbirds. Five years before, he had caught a partridge on the night of the first of September; he thought that probably the shooting had driven it to sea. He had also captured a young seal by descending from his perch in the lighthouse, and placing a bag in front of him as he slept. "He poked him up behind with a stick, and in he went."

A foreign ship once struck on the end of the rock in broad daylight. The crew, twelve men, leaped on shore; the vessel drifted about three miles and sank. On being asked how they had fed so many men, the keepers replied that they always had six months' provisions when they entered upon their charge.

As to the old lighthouse. It was ascended by a rope ladder. The piles, though they had stood for eighty years, looked exceedingly insecure; they were set upright in the rock, with a few props on one side to resist the greatest pressure of the waves. The upper part consisted of a kind of platform, on which were placed those provisions and stores not easily damaged by water. Above it rose a wooden barrack—an octagonal cabin, in which the workmen employed about the new building slept in berths, like those of a passenger steamer; on the next story was entered, through a trap-door, the keepers' sleeping-room and kitchen, both in one; and topmost of all was the lantern. In heavy

THE SMALLS LIGHTHOUSE.

weather, when the sea was dashing about the lower room, the workmen and keepers congregated in the upper. The whole structure, in such weather, trembled and swayed to and fro, and had been known to lean nine inches from the perpendicular.

As to the new lighthouse. The Commissioners proceeded to examine its mode of construction. They observed that the stones were all prepared and carefully fitted on shore—that, in fact, the lighthouse was actually built there. " Each stone has a square hollow on each edge, and a square hole in the centre; when set in its place, a wedge of slate, called 'a joggle,' fits into the square opening formed by joining the two upper stones. The joint is placed exactly over the centre of the under stone, into which the joggle is wedged before the two upper stones are placed. The result is, that each set of three stones is fastened together by a fourth, which acts as a pin to keep the tiers from sliding on each other. The base of the building is solid. Two iron cranes slide up an iron pillar in the middle, and are fixed by pins at the required position as the work advances. The two are used together, so as to obviate any inequality of strain."

The building was completed in 1861. It is coloured externally with red and white horizontal bands, and is one hundred and forty-one feet in height. It shows a white light for fifty-four seconds, eclipsed for two seconds; light for two, re-eclipsed for two. A fog-signal (explosive) gives a report every five minutes.

Rounding the southern coast of Wales, we observe on *St. Ann's Head,* Milford Haven—the finest natural harbour of Great Britain—two lighthouses of substantial construction. The higher is a circular tower, seventy-five feet in height, erected in 1714; the lower, lighted for the first time in the same year, two hundred yards to the south-east, is octagonal in shape, and only forty-two feet high. Both exhibit a fixed white light, which is visible for twenty miles from the higher (with a red sector in a westerly direction) and eighteen miles from the lower station. These lighthouses are easy of access from Milford, and permission to view them is easily obtained. A fog-siren is sounded in thick or foggy weather. A coal-fire light was exhibited on this point in 1714; replaced by oil in June 1800.

Caldy Island, in lat. 51° 37′ 56″, shows an occulting white light, visible for twenty-seven seconds, eclipsed for three, with a red sector, visible twenty miles, from a well-built circular tower of limestone, fifty-six feet high. It dates from 1829; was designed by Mr. Nelson; and cost, with adjoining buildings, £3,380, 11s. 7d.

Passing the lights of Tenby, Saundersfoot, Barre, and Llanelly, and the *Helwick* lightship (which displays a revolving white light), we arrive at the *Mumbles,* off Swansea, and the mouth of the Tawe, in the same latitude as London. This important position is well marked by a lighthouse, erected in 1798, on the outer

islet—a white octagonal structure, which, at a height
of one hundred and fourteen feet above high-water
mark, extends over a range of fifteen miles a brilliant
white light.

We now sight in succession the *Scarweather* lightship, in lat. 51° 26′ 55″, with its revolving eye of red,
the lights of Porthcawl, and the two upon the *Nash
Point*, in lat. 51° 24′. Both these lighthouses were
designed by the late James Walker. They are circular
stone towers, with stone lantern gallery, and massive
walls. The eastern, or high, measures one hundred and
thirteen feet, and the western, or low, sixty-seven feet,
from base to vane. Their cost (with adjoining buildings) was £5,796, 14s. 1d. Each shows a fixed white
light (with a red sector in the eastern), visible nineteen
and seventeen miles respectively.

Passing the *Breaksea* lightship (with revolving white
light), we come to *Flatholm Island*, off Cardiff, where,
in 1737, a coal-fire was kindled for the guidance of
mariners ; replaced by an oil light in September 1820.
The present light-tower, of stone, built in 1737, was
restored in 1881, and exhibits an occulting white light,
with white and red sectors. The lights of busy Cardiff
and of ancient Usk recede behind us as we approach
the *English and Welsh Grounds* lightship, lat. 51° 23′
30″, the revolving white light of which shows an
intense glow every fifteen seconds. On the east side
of the mouth of the Avon stands a strong octagonal
tower of brick, designed by Mr. Walker, and completed

on May 25th, 1840, which exhibits a white light, with red and green sectors; bright for twenty-seven seconds and eclipsed for three. Passing the lights on Portishead and Clevedon piers, we descry the two lighthouses at Burnham, on the east side of the estuary of the Parret; one, of brick, ninety-nine feet high, the other, of oak, thirty-six feet; both designed by Mr. J. Nelson, and dating from 1832; the high light, white, occulting, and the low light, fixed, white and red. Leaving behind us the lights at Watchet Harbour and on Lantern Hill, Ilfracombe, we perceive on *Bull Point* (lat. 51° 11′ 5″) a circular light-tower, built in 1879, with an upper light of the group-flashing type, and a low red light marking the ominous Morte Stone. On *Braunton Barrows*, north side of the river, stands an octagonal structure of oak, on foundation of concrete; eighty-six feet from base to vane; erected in 1820 from Mr. Nelson's plans; and displaying a fixed white light, visible for fourteen miles. Also, a small timber light-room, on piles, three hundred and eleven yards distant, erected at the same time, but rebuilt in 1832; shows a fixed light over eleven miles. This part of the Devonshire coast is very dangerous. In his "Westward Ho!" Kingsley describes it as "a waste and howling wilderness of rock and roller, barren to the fisherman and hopeless to the shipwrecked mariner."

Off the mouth of the Bristol Channel lies *Lundy Island*—measuring about three miles in length by one mile in width, with nearly twenty miles of coast-line—

a rugged mass of gray granite, so guarded by insulated rocks that, according to a popular local saying, " there is no entrance but for friends." Rabbits abound here, and the cliffs are white with the wings of innumerable sea-birds, whose screams fill the air and are repeated by every echo. Here it was that Sir Lewis Stukely, the betrayer of Sir Walter Ralegh, fled from the scorn of men, and died of remorse and solitude, a maniac, in 1620. On the rocky summit of Chapel Hill are situated the granite lighthouse tower, ninety-six feet high, and the keepers' dwelling-house. Two lights have been displayed here since 21st February 1820 ; the upper a white flashing light, every minute, and seventy feet lower a fixed white light ; both having a fair-weather range of about thirty miles, though it is said they have been seen at a distance of forty-five miles. The buildings were designed by Mr. David Alexander, and cost £10,276, 19s. 11d. The lantern, twenty-eight feet high and one and a half foot diameter, cost £1,902, 18s. 4d. Some improvements were effected in 1889.

Keeping along the precipitous Cornish coast, and having taken leave of Devonshire at *Hartland* Point (lat. 51° 1′ 24″), where a handsome new lighthouse has recently been erected, with a revolving white and red light, we arrive at *Trevose Head*, in lat. 50° 32′ 55″ (about two and a half miles from Padstow), crowned by a strong white tower of stone, eighty-seven feet high, and two hundred and four feet above the sea, exhibiting a white occulting light (three occultations every minute), which has a sea-range of twenty miles.

Off *Godrevy Island* the fine iron screw-steamer *Nile* was totally wrecked on the 30th of November 1854, and all on board, crew and passengers, perished. This and other similar calamities led to the erection (from Mr. James Walker's designs) of the present lighthouse, which was first lighted on March 1st, 1859, and displays a couple of lights—a white light, flashing ten seconds, and a fixed red light, both visible for fifteen miles. Cost, with adjoining buildings, £7,331, 4s. 5d.

The light-tower is built of rubble stone bedded in mortar. Octagonal in shape, and eighty-six feet high, it is planted on a rock of considerable size, where numerous wild plants relieve with their greenery the prevailing aspect of desolation. In the summer season it is a favourite resort of excursionists from Penzance and St. Ives, as many as a thousand persons visiting it on a Whit Monday.

This lighthouse indicates the position of the dangerous reef called "The Stones," near St. Ives. It was designed by Mr. Walker, engineer to the Trinity House. Cost £7,082, 15s. 7d.

Passing the lights of Padstow, Hayle, and St. Ives, we see before us the *Seven Stones* lightship, and find that we have completed our survey of the coast of England and Scotland.

We proceed to inspect that of Ireland, beginning at Fastnet, on the south, in lat. 51° 23′ 18″.

CHAPTER VI.

ABOUT four miles and a quarter south-west from Cape Clear, the southernmost point of Ireland, in lat. 51° 23′ 18″ and long. 9° 36′ 25″, stands the *Fastnet* (or *Fastness*) *Rock*, with its beautiful lighthouse.

From certain effects of lights, and more particularly when the sun is in its vernal equinox, this rock, when seen from the shores of Cape Clear or the adjacent islands and headlands, presents "a peculiarly spectral appearance," easily mistaken by strangers for that of a large ship under sail; and this appearance may possibly, as Mr. Sloane suggests, have originated the old fable that every May morning the rock sets sail, cruises round the Darsey Island, visits the Bull, Cow, and Calf Rocks, its kith and kin, and then settles down again in its time-old position.

It was long credited with the remarkable property of being just nine miles from everywhere. This mistake was easily exposed by the Ordnance Surveyors;

but credulous folk are still ready to maintain that such
was originally the case, but that the last time the rock
returned from its cruise, it made some mistake in resuming its former position. Another tradition exists
among the people of West Cork, that the Fastnet Rock
was picked out of Mount Gabriel, where a lake is
pointed out as filling the cavity caused by its removal;
and some are convinced that articles thrown into this
lake will duly reappear by some underground and
undersea passage on the Fastnet Rock. Yet again, it
is said that the remarkable gap or breach in Mount
Gabriel was caused by the devil's voracity in biting a
mouthful out of it, which, finding it unpalatable, he
dropped where, in later ages, it has been known by the
name of the Fastnet Rock.

Apart from this traditional glamour, the rock is
interesting from its picturesque character, its solitari-
ness, and because it is the last bit of the old country
seen by the emigrant who is bound for other shores—
whence it is often called *Tear Erin*, "the Tear-drop of
Ireland," summoning up the tears of those who long-
ingly "look back to that dear isle they are leaving."

The lighthouse on Cape Clear was so frequently
obscured by mist and fog as to be of little service
to shipping; and it was decided, therefore, in 1848, to
erect one on the Fastnet Rock. The design was furnished by the late Mr. George Halpin, and consists of
a tower composed of a casing of cast-iron plates, with
flanges and stiffening ribs, the lower story of which is

partially filled in with masonry, leaving space for a
coal-vault, and the other stories lined with brickwork;
the floors are of cast-iron plates laid on radially dis-
posed girders, which unite and rest on a central hollow
column, and bind the tower at each story. Of these
there are five, twelve feet high, measured from floor to
floor, the internal diameter of the tower being twelve
feet also. The height from base to gallery is sixty-
three feet nine inches, above which rises a well-
proportioned lantern, uniting apparent lightness with
the requisite strength. "It is hardly necessary to observe that the management of the different portions
of this tower to meet the heavy shocks of wind and
sea was an effort of no ordinary engineering skill; and
although differences of opinion may exist as to the
fitness of such structures for lighthouse purposes, there
is, perhaps, no other method by which a lighthouse
could be placed in such a situation so speedily or
economically."

The cast-iron plates were all landed on the rock by
June 1849; and thenceforward the laborious and
difficult undertaking was prosecuted with so much
energy that on the first night of January 1854 the
lamp of the Fastnet was able to be lighted, and its
bright flashes shot across six leagues of the great
Atlantic. The apparatus is dioptric, of the first order
of Fresnel, revolving once a minute. Its focal plane
is one hundred and forty-eight feet above the sea; but
the building itself, which is painted white, with a broad

horizontal belt of red midway, is ninety-two feet in
height from base to vane. In 1867-9 it was cased round
its base with metal plates for twenty-four feet up, and
between this casing and the outside of the tower rubble
masonry grouted in cement was filled in solidly. The

FASTNET ROCK LIGHTHOUSE.

dwellings of the keepers and their store-houses are,
like the tower, of cast-iron. Cost £18,947, 15s. 11d.

In this vicinity we meet with another important
lighthouse—one of the finest structures of the kind in

the world—that which crowns the summit of *Gally Head*, a precipitous cliff near Cape Clear, in lat. 51° **31′ 50″.** It was erected, or at least completed, in 1878, from the designs of Mr. John S. Sloane, C.E., late engineer **to** the Irish Lighthouses Commission. Besides the handsome circular light-tower, sixty-eight feet in height—with the focal plane of its lantern one hundred and seventy-four feet above the sea—which stands enclosed within a substantial stone wall, there are dwellings for the keepers on **an** exceptionally complete scale, each with its separate approach and garden, also engine-house and gasometers; the whole walled in very neatly, and covering a very considerable area.

If the Gally Head Lighthouse be, as is asserted, unequalled in its appointments, it is probably unequalled in the power of its illuminating apparatus, which is constructed on the system of Mr. John **R.** Wigham of Dublin, **so well known** from his services in connection with lighthouse illuminants. Briefly speaking, its light may be described as proceeding from a quadriform arrangement of gas-burners, used without chimney glasses or any interposing medium. Each burner has an illuminating power of one thousand two hundred and fifty-three candles; and the great beam **of** light yielded by the whole combination is about thirteen feet high by three feet wide. This beam or luminous column reaches the mariner every minute in the form of a **group** of six or seven flashes, lasting for about sixteen seconds, and followed by an interval of obscurity of forty-four seconds—the alternating flashes and darkness being succeeded by alternate extinction and reignition of the gas through clock-work machinery, so that about one half the consumption of gas is saved, while the effect is exceedingly distinctive.

The novel feature of Mr. Wigham's ingenious mode of lighting is to be found in the superposing of lights and lenses. In his quadriform apparatus at Gally Head, thirty-two lenses are arranged in four tiers, eight in each tier. Each lens is of the size of the first-order lens. A powerful gas-burner, with sixty-eight jets, invented by Mr. Wigham, is set in the focus of each tier, so that four burners are placed one above the other. The lenses being contiguous, all the jets blend at the distance of a few yards, forming the great pillar of light already spoken of, the illuminating power of which has been computed equal to nearly one *million sperm candles*. It is to be noted that this vast illuminating power is, by a well-conceived arrangement, placed so completely under the lighthouse-keeper's control that only one-fourth of it is applied in clear weather, the other three parts being reserved for application as the weather thickens. This Gally Head light is the largest in the world.

The method adopted for extinguishing and relighting the gas, so as to produce the flashes, is of happy ingenuity, a motion being obtained from the clock-work on the valves, similar to the Cornish or bevelled valve.

the bevel being underneath, but the surface of metal
to metal is horizontal, truly termed. This bevel is
framed to stop an instantaneous rush of gas ; and the
means of lighting comes from a small pipe connected
with a single jet, in which a small portion of flame is
continually burning, so that immediately the valve is
opened the small jet ignites the gas issuing from the
eight-and-sixty jets. The other three burners are
worked in the same manner. This lighthouse has
three screens, which cut off the light east-northerly
and west by north-westerly. The flue to the chimney
is inside the sector, and therefore does not obstruct the
light. For production of gas five retorts have been
provided, though only one is required to produce the
gas for consumption. There are two gas-holders for
storage. In the event of any mishap occurring to the
gas, provision is made by which the usual Trinity oil-
lamp can be substituted in less than thirty minutes.

  The light at Gally Head has been named by Mr.
Wigham "the group-flashing light," the flashes being
produced, as we have seen, by a process quite new in
lighthouse illumination. The occultation of a fixed
light, as first illustrated by Mr. Babbage, may be
effected by causing opaque screens to close at certain
intervals automatically round the light; or the occultation may be effected, as at Wicklow Head, by the
lowering at given intervals of a gas flame. The arrangement at Gally Head is totally different. Instead
of allowing the broad beam of light to pass continu-

ously, as in the ordinary revolving light, a simple
automatic apparatus cuts it up into a series of flashes,
sufficient in number to insure that they can never
wholly escape the mariner's attention, and in each of
which a flame of great power is brought into play.

On *Kinsale Old Head* a very fine stone tower, one
hundred feet high, painted white, with two red belts,
designed by the late George Halpin, was erected and

KINSALE LIGHTHOUSE.

lighted in 1853. The light is of the first-order dioptric, white, with red sector (in the direction of the
Horn Rock), and commanding a range of twenty-one
miles. This light is two hundred and thirty-six feet
above the sea. The present building would seem to
be the third of its kind on the Old Head. The remains
of a tower lighted in 1805 still stand a short distance
to the north. An earlier structure, dating from 1683,
was either at or near to Barry Oge's Castle, an ancient

fortification designed to separate the promontory from the mainland, **of** which the ruins are **in** tolerable preservation. Cost of lighthouse, £10,584.

A lightship, **with fixed red** light, is moored rather more than a mile south of the *Daunt Rock.*

Passing the lights **of** Cork, we come to *Ballycottin Island,* in lat. 51° 49′ 30″, where a circular stone tower, fifty feet high, enclosed within white walls, was first lighted in June 1851. The lighting apparatus, the focal plane of **which is** one hundred and ninety-five feet above the sea, shows a brilliant white flash every ten seconds. In a belfry close at hand a bell is tolled by machinery in foggy weather. Designed by Mr. Halpin. Cost £11,746, 15s. 5d.

At *Youghal* the circular light-tower of cut stone, forty-three feet high, designed by Mr. Halpin, was lighted in 1852, at a cost of £4,679, 6s. 5d. The ruins of the ancient nunnery of St. Anne's, and its tower lighthouse, were removed **to** make way for it; and all that now remains of interest is the tradition that in the nunnery garden, when it belonged to **Sir Walter** Ralegh, potatoes were first planted in Ireland. The lantern shows a fixed white light, while a red light is exhibited from a window. Visible for six miles.

On the south side of *Minehead,* on the coast of Waterford, lat. 51° 59′ 33″, a tower-lighthouse of solid masonry, designed by Mr. Halpin, and sixty-eight feet high, on a cliff two hundred and twenty feet above the **sea, forms an important guide for Dungarvan Bay. Its**

light, visible for twenty-one miles, is supplied by **Mr. Wigham's gas** apparatus; **is** shown for fifty **and eclipsed for ten** seconds every **minute. Cost** £9,799, **19s. 7d.**

On *Ballinacourty Point*, **Dungarvan, a** circular **tower of** limestone, **designed by Mr. G. Halpin, forty-four** feet high, **was** lighted **in 1858. The light** shows *red* **in the** direction of Carrickapane Rocks, *green* in the direction of Ballinacourty **Rocks, and** *white* **in** all other directions **in which it is visible.** Cost **£6,737, 6s.**

Among the public records preserved in Dublin Castle **is to be seen a letter from one** Robert **Reading,** dated **September 12th, 1671,** applying for a pension **of** £500 **per annum, out of the** "concordata" for light-houses built **by him according to** letters patent. These lighthouses were **at Howth,** on Magee Island, at the Old Head **of Kinsale, near Barry Oge's** Castle in Kinsale, **and at Hook Point.** *Hook Tower* is situated **on Hook** Point, **east side of the entrance to** Water-ford **Harbour** (*Vader Fiord*, "the Great Haven"— anciently **called** *Cuan-na-Grioth*, **or** "Harbour of **the Sun"), and** presents **both from** land **and sea** a remarkable appearance. **It** is, perhaps, the old-est lighthouse **tower in the** United Kingdom. The lower portion **is,** horizontally, **an** irregular ellipse **of** forty-two feet, rising vertically **to a** height of eighty **feet;** internally divided into three great central vaults, rudely but substantially built **of** massive stone-work, chiefly rubble, in courses. **In** the lower vault, now used as the coal-store, is preserved the guard-bed used by the military when the tower was garrisoned in 1798. The middle and upper vaults have been converted into rooms for the keepers.

In 1864 this was changed from a catoptric to a dioptric station. The old lantern, erected thirty years before, to supersede the original one of 1791, consisted of a murette or blocking of oak covered with sheet copper, raised on the top of a secondary tower, nineteen feet in diameter and about thirty feet in height, with a vaulted roof of great strength. The principal approach was by an outer stair for about half the height, and then by a stair in the wall eighteen inches wide. The oil was brought up through a trap in the lantern floor. In removing the vaulted top of the turret a large bed of cinders was discovered, the remains of the original coal-fires used to illuminate the beacon prior to the erection of the first lantern in 1791.

In 1864 the tower was raised, and a new lantern constructed. It is now one hundred and fifteen feet high, and the focal plane of the lantern one hundred and fifty feet above the sea. Mr. Wigham's gas apparatus is in use here, furnishing a magnificent fixed white light, visible for sixteen miles.

Passing Waterford, we arrive off the Saltee islands, close to the southernmost of which is moored the *Coningbeg* or *Saltees* lightship, in lat. 52° 2′ 25″, and one of the most exposed positions in the kingdom. She is coloured black, with a white stripe, to distinguish her from the floating lights on the English coast, which are red. She has a white light, showing three flashes in quick succession every minute; and a fog-siren, which gives a blast of two and a half seconds, repeated after an interval of twenty-five seconds, and followed by ninety seconds of silence. First stationed in 1824. The *Barrels Rock* is likewise indicated by a lightship, moored (in 1880) at a distance of two and a quarter miles south-south-west, which shows two flashes of red light in quick succession.

On the *Tuskar* **Rock**, in lat. 52° 12′, stands a stately circular tower of granite, erected in 1815, from Mr. Halpin's designs, and raised to its present height—one hundred and thirteen feet—in 1885. Its foundation is below high-water mark. While the works were in progress, in the winter of 1812–13, a tremendous storm swept over the rock and carried away a considerable portion of the foundation, drowning and wounding many of the workmen. The lantern displays a red and white light alternately, revolving every minute; and in foggy weather a guncotton-powder charge is exploded a little above the lantern at intervals of five minutes. Similar charges are used for danger-signals. Original cost, £35,887, 17s.

Between Rosslare and Wicklow are stationed the *Lucifer Shoals*, the *Blackwater* **Bank**, and the *Arklow Bank* lightships. Beyond Wicklow a lightship is moored off the **Codling Bank**. Dublin Bay is amply lighted, from the Kish lightship, in lat. 53° 19′ 25″, to the pier-head at Howth, in lat. 53° 23′ 35″; but we must limit our notice to the *Baily* light-tower, situated on the south-east point of the Howth peninsula, which, from its vicinity to Dublin, and the picturesqueness of its position, enjoys a considerable reputation.

The present Baily light is supposed to be the third which has stood on this peninsula, which is about three miles long by two miles broad, and rises to a height of five hundred and sixty-three feet. The former lights were set up on nearly the most elevated point, and consisted, it would seem, of a large floor or hearth of granite, slightly hollowed, on which a coal-fire was kept blazing. Subsequently (about 1671) a small tower was built, but no record remains of its mode of illumination. The present structure, a stone tower, forty-two feet from base to vane, was designed by Mr. Halpin, and erected in 1813, on the site of the ancient fortress of Dun Criffan. Its focal plane is one hundred and thirty-four feet above the sea. It is lighted by Mr. Wigham's patent gas apparatus, which throws a fixed white light over a range of fifteen miles. A fog-siren is stationed here, and worked by a caloric or hot-air engine.

On the top of the outer crag, called Rockabill, in lat. 53° 35′ 47″, Mr. Sloane erected, in 1860, a very fine circular light-tower of gray stone, one hundred and five feet in height, illuminated by white and red sectors, which show flashes every twelve seconds, eighteen miles, on Mr. Wigham's system of quadriform gas-burners.

Dundalk channel is indicated by an iron octagonal lighthouse, on screw-piles, built in 1854–5, from Mr. Halpin's designs, and carrying a small dioptric flashing light of fourth order, with white and red sectors. In foggy weather a bell is rung six times a minute. Cost £6,068, 3s. 5d.

The *Haulbowline Rock,* Carlingford, is surmounted by a tower of very beautiful design, built in 1819–26, and lighted in 1823. Cost £28,396, 17s. 3d. The original Carlingford lighthouse, erected by Mr. Rogers, stood at Cromfield Point, where the keepers' residences now are; its remains were removed in 1864. The present, designed by Mr. Halpin, is a shaft of granite painted white, with an elevation of one hundred and eleven feet from base to vane. At high water the centre of its lantern rises one hundred and four feet above the sea; so that the tower is submerged for seven feet from its base. The light is catoptric, and visible for fifteen miles.

On *St. John's Point,* Dundrum Bay, where the *Great Britain* was stranded in 1849, a lighthouse tower of cut stone, seventy-three feet, was erected in 1844, from Mr. G. Halpin's designs, and is furnished with a dioptric light of second order, illuminated by gas. Cost £11,091, 1s.

The light-tower built by Mr. Rogers on the *South Rock,* locally called the Kilwarlin, is no longer used. Since 1877 its position has been indicated by a lightship, which carries a white light, revolving in a minute and a half. A lightship is also stationed about a mile from the *Skulmartin Rock*.

Off the south-east side of the mouth of Belfast Lough lies *Mew Island*, or the *Smaller Copeland*, in lat. 54° 41′ 50″. Here stood one of the most ancient of the Irish lights, described as of a square form, with walls seven feet thick, and seventy feet high to the lantern. "It consists of three stories, of which the lower and second are laid with beams and boarded, but the third is arched and covered with large flagstones seven or eight feet in length. In the middle of the house is erected a round tower, on which the grate is fixed on a thick iron spindle. Scotland supplies it with coals, of which, on a windy night, it consumes a ton and a half, burning from evening till daylight, both winter and summer." The present tower, fifty-eight feet high, was built by Mr. Halpin (1813–16), and its focal plane is one hundred and thirty-two feet above high water. Its light was improved in 1884, and is now a group-flashing one. The flashes occupy about twenty-two seconds; and between each successive group occurs an interval of thirty-eight seconds. A fog-siren is stationed here. Original cost of lighthouse, £9,051, 17s. 6d.

Crossing Belfast Bay, we come to the mouth of Lough Larne, where, on *Farres Point*, in lat. 54° 51′ 7″, was erected in 1839 a lighthouse of cut stone, fifty feet high, designed by Mr. Halpin. It is a catoptric light, fixed, bright, with a red sector in the direction of the Hunter Rock. Cost £7,358, 18s. 3d.

The *Maidens Rocks*, East and West, locally known as the Huillans **and** Ullans, lie about half a mile apart, and eight miles from the shore. The East lighthouse, which is sixty-eight feet high from base to vane, and the West, which is seventy-six **feet, were** both designed by the late **Mr.** Halpin, and lighted **for the** first time in **1829.** **They** are built of white stone, with a dark-red **belt** painted round each. The light **in** each is catoptric, **of the first** order, **white,** and fixed, with a **range of thirteen and** fourteen miles. In 1887 a sixth-**order dioptric red** light was shown from a window **of East** Maidens lighthouse, **to** mark the Russell and Highland **rocks.** Cost of each, £18,526.

On *Altacorry Head,* **the** north-east point **of** Rathlin Island—a crescent-shaped basaltic mass, once **the** asylum of Robert Bruce—a lighthouse tower of gray stone, eighty-eight feet high, **with the** focal plane **two hundred** and forty feet above **high water,** was erected in 1856 by **Mr. Halpin.** It has a dark-red belt under the gallery. **The illuminating** apparatus con-**sists** of two powerful dioptric lights of the **first order**— the upper, intermittent, showing a bright light for fifty seconds and being eclipsed for **ten;** **and** the lower light, about twenty-five **feet from** the rock, **being a** fixed white light. **The** upper is visible **for twenty-one** miles. During foggy weather **a** gun **is fired every** quarter **of an hour.** Cost £9,649.

Steaming **past the** columnar basaltic cliffs **of** the Antrim **coast, and the** marvels of the Giants' Causeway,

we arrive **at the mouth** of Lough Foyle, where at
Inishowen, **or** Owen's Island, **two** lighthouses upon
Dunagree **Point** (lat. **55° 13′ 38″**) mark out the entrance channel. Both are upon exactly the same level,
but the eastern is forty-nine feet high, and the western
seventy-four feet (having been raised twenty-five feet
in 1870). Both are built **of** stone, date from 1837,
and were designed by the late Mr. Halpin, at a cost of
£8,749, 14s. 3d. for the western, **and** £9,104, 14s. **2d.**
for the eastern. The eastern has a fixed bright light;
the western a red **sector.** They are visible for fifteen
miles. The river Foyle falls into the lough, and on its
banks, about four miles inland, clusters **the** historic
city of Londonderry.

Continuing our voyage along the **north** coast, we
come **to** the small island of *Inishtrahull,* on the
northernmost point of Donegal, in lat. 55° 25′ 55″, lying
about a mile from the mainland, where, in 1812, a
lighthouse tower of good stone-work, forty-five feet
high, was erected by the late Mr. G. Halpin, at a cost
of £10,850, 8s. 4d. It is furnished with a dioptric
light of the first order, which revolves every thirty
seconds, and is visible for eighteen **miles.** Height
above high water, one hundred and eighty-one feet.

The two boundaries **of** Lough Swilly, *Dunree Head*
and *Fanad* (or *Fannet*) *Point,* are both lighted—the
former by a fixed white light shown from a dwelling-
house; the latter from a small circular tower, twenty-
six feet high, **erected in 1816 from** Mr. Halpin's

designs, at a cost of £5,756, 1s. 10d., and in 1886 equipped with an illuminating apparatus of the second-order dioptric, which gives an occulting light at half-minute intervals, white seaward, and red from north-west to land. The focal plane is one hundred and twenty-seven feet above the sea. A fixed white light is shown at an elevation of seventy-two feet.

The tower on *Tory Island*, so named from the *tors* or high cliffs on its eastern side, was designed by the late Mr. Halpin, and erected in 1832. It is of circular form, and built of cut stone; eighty-seven feet high; and cost £16,750, 0s. 7d. It exhibits a fixed white light, of the first-order dioptric, at an elevation of one hundred and twenty-five feet above high-water mark. Tory Island is rich in relics of antiquity and natural curiosities. But the sweeping winds from the Atlantic strip it of vegetation; and tradition records that a boat's crew of its inhabitants, having been driven by stress of weather into Ardee Bay, seven miles distant, were struck with wonder on landing to see the green trees and leafy bushes, carrying away twigs and branches to exhibit on their return to their treeless home as memorials of an earthly paradise.

On *Rinrawros Point*, the north-west extremity of Aranmore Island, rises a stone light-tower, seventy-six feet high, with its focal plane two hundred and thirty-three feet above high-water mark. It was erected in 1864 from the designs of Mr. Halpin, and is equipped with a dioptric light of the first order, which shows a group of flashes in rapid succession (1887). A fog-siren has been established here. The old lighthouse, the date of which is not known, was disused about 1829, and the greater portion of its masonry was utilized in the construction of the new tower and the keepers' dwellings.

On the west side of the island of *Rathlin O'Birne*, which lies off Teelin Head, on the north-west coast of Donegal, a tower-lighthouse of solid masonry, sixty-five feet from base to vane, was erected by Mr. Halpin in 1856, at a cost of £17,140, 0s. 7d. Its light is a first-order fixed catoptric, visible for sixteen miles, white to sea-ward and red towards the mainland.

The harbour of Killybegs is protected by two lights: on *St. John's Point*, north side of Donegal Bay, in lat. 54° 34′ 8″, Mr. Halpin, in 1831, erected a circular stone tower, forty-seven feet high, with a fixed white light, ninety-eight feet above the sea-level. Cost £9,606, 7s. 4d. On *Rotten Island*, lat. 54° 36′ 21″, the building is also of stone, measures forty-seven feet from base to vane, and exhibits a fixed white light, leading to the passage from seawards, and the inner harbour channel, clear of the harbour rocks. Designed by Halpin; cost £8,867, 15s. 11d.

· We are brought in our south-westerly course to Sligo Bay, where the *Black Rock*, on the south side of the entrance, exhibits a fixed white light from an elevation of ninety-four feet. The lower portion of the tower is

solid masonry, and was built for a beacon, in 1822, by
Mr. Halpin, but completed as a light-tower in 1835.
The keepers' residences are of cast-iron. Cost £9,921.
Two towers, each forty-three feet high, but about four
hundred and five feet apart, were erected on *Oyster
Island*, within the entrance to Sligo Harbour, in 1837
and 1857, by Mr. Halpin. Cost £4,600 each. Both
are illuminated by fixed white lights. It has been
proposed to discontinue these lights; to remove the
north lighthouse one hundred and eighty yards westward, and exhibit from it a red sector guarding Bungar
Bank, and a white light to indicate the entrance channel of the river.

On *Gubacashel Point*, which forms the west boundary of Broadhaven, the tower (of dressed stone, with
solid walls) is fifty feet high, and eighty-seven feet
above high water, with a fixed white light. Cost
£5,702, 1s. 8d.

Eagle Island lies off the mouth of the Mullet, in lat.
54° 17', and long. 10° 5' (in the county of Mayo), where
beats the heaviest sea that is known on the coast of
Ireland. There are two circular stone lighthouses upon
it—one, westward, is forty-four feet high, and two
hundred and twenty feet above the sea, with a fixed
bright light; the other, one hundred and thirty-two
yards to the west, is eighty-seven feet high, with a fixed
bright light, two hundred and twenty feet above the
sea. Both were erected in 1835 from Mr. Halpin's
designs, at a cost of £18,111, 4s. each. Their lanterns

have more than once been injured by the sea—a mishap unknown at any other station on the coast.

Not less wild and difficult of access is the *Black Rock* of Mayo, where a light of the dioptric first order, holophotal, white towards the sea and red towards the land, showing a flash every thirty seconds, was first displayed in 1864. The buildings and tower are two hundred and thirty feet above the sea; the tower fifty feet in height.

Passing *Blacksod Point* (a fixed light, with white and red sectors, in a reddish-gray tower), we come to *Clare Island*, one of the ancient Irish lighting-stations, westward of Clew Bay, in lat. 53° 49′ 30″. A light was first shown here in 1806. The present tower is thirty-nine feet high, and planted in a boldly romantic position, three hundred feet above the sea. Cost £9,297, 13s. 6d. At *Inishgort*, on the south point of the island, showing the entrance to Westport, another lighthouse was erected twenty-one years later, from Mr. Halpin's designs. While the former has a range of twenty-seven miles, the latter is limited to ten miles, and stands upon a low shore, only ten feet above high water. The tower is twenty-six feet from base to vane. Cost £3,180, 0s. 10d.

As we approach Galway Bay, we sight the wild-looking cliffs of *Slyne Head* (the ancient "Ceame Leame"). On Illaunimmul, the outer islet, a noble tower of stone, seventy-nine feet high, with keepers' dwellings, was designed by the late Mr. Halpin. A

similar tower, of nearly the same height, one hundred
and forty-two yards to the south-west, was built at
the same time, and both were lighted on the same
night in 1836. The former is a revolving light, showing one red and two white faces alternately every two
minutes; the other a fixed white light. Cost £20,823
each.

Galway Bay is lighted by a tower, one hundred and
one feet high, erected in 1857 (from Mr. Halpin's designs) on the west point of *Eeragh Island*, in lat. 53°
8′ 55″; a light on *Straw Island*, at the entrance to
Killeany Bay; and lights on *Inisheer* and *Mutton
Islands*. On Mutton Island, the tower, built in 1819
by Mr. Halpin, is only thirty-four feet high, and the
shore so low that at high water the light is barely
thirty-three feet above the sea. Cost £4,020, 3s. 5d.
On the south point of Inisheer Island, one of the
chain of the Aran Islands, is a fixed white light
(with a small red sector over the Finis Rock) in one
of Mr. Halpin's lofty and graceful edifices, a tower
of limestone measuring one hundred and twelve feet
from base to vane, dating from 1857. Cost £14,252,
2s. 4d.

On the north side of the entrance to the beautiful
river Shannon projects the westernmost point of
County Clare, *Loop* (or *Leap*) *Head*, where a light
was first exhibited in 1802. The present tower, of
cut stone, painted white, dating from 1853, and designed by Halpin, is seventy-five feet high, and

cost £6,684, 11s. 5d.; and its occulting white light, visible twenty-two miles, is placed two hundred and seventy-seven feet above high-water mark. A legend blooms in this picturesque spot. Cuchullin, a great chief of the Red Branch Knights, was hotly pursued by his enemies to this remote quarter, where they gradually fell off, weary and spent, until only one remained, and that one a woman. Seeing a lofty rock before him, separated from the mainland by a gulf of heaving waters, he leaped across. The woman courageously followed. Bracing up all his energies, he leaped back again. The woman made a similar effort, but failed, and fell into the boiling sea.

Passing *Kilcradan Point*, where Mr. Halpin, in 1824, built a stone lighthouse, circular, forty-three feet high, we perceive a fixed white light, with red sector, upon *Scattery Island*, where in 1872 the tower and keepers' residence were erected, close to the fort, by the late Mr. Sloane. In a metrical life of St. Senanus, preserved among the *Acta Sanctorum Hiberniæ*, we are told that this misogynist fled from the face of woman to Scattery Island, and would admit no one of the hated sex to visit his retreat, not even a sister saint, St. Cannera, whom an angel had taken thither to introduce to him. According to his poetical biographer, he replied in Latin verse:—

" Cui, Præsul, quod fœminis
Commune est cum monachis?
Nec te nec ullam aliam
Admittemus in insulam."

The legend supplies the motive of one of Moore's Irish melodies, "St. Senanus and the Lady":—

> " Oh ! haste and leave this sacred isle,
>   Unholy bark, ere morning smile ;
> For on thy deck, though dark it be,
>   A female form I see ;
> And I have sworn this sainted sod
> Shall ne'er by woman's feet be trod."

The lights on *Tarbert Rock* and *Beeves Rock* are both in the lower estuary of the Shannon. Both were designed by Mr. Halpin, and erected in 1834 and 1854 respectively. The Tarbert lighthouse is fifty-four feet high, of white stone, romantically situated, and approached by a light and graceful metal bridge. Cost £10,112, 15s. 3d. The Beeves lighthouse is of cut stone, and about seventy-four feet from base to vane. Cost £9,494, 8s. 1d.

The river-channel to Limerick is marked out by fourteen lights, exhibited chiefly on poles and perches. To guide vessels to the entrance to Tralee Bay, a circular lighthouse of blue stone was erected in 1854 upon *Little Samphire Island*. It is forty-two feet high, was designed by Halpin, and cost £7,006, 17s. 1d.

Tearaght, one of the Blasket Islands, in lat. 52° 4' 30", is the most westerly inhabited point in Europe. A white light, showing two flashes in quick succession, was set up here in 1883. The tower, however, was erected in 1864-70 from the designs of Mr. Sloane ; is fifty-seven feet from base to vane, while its light is two hundred and seventy-five feet above the level of high water. The island or rock is worth visiting for the romantic character of its scenery. It is divided by the sea into two portions, which are connected by a natural arch or bridge, and rises at its summit to an elevation of six hundred feet.

Cromwell **Point,** Valentia, derives its name from an ancient fortalice which stood there, the old walls of which can still be traced in the lighthouse enclosure. The tower, designed by Mr. Halpin, is forty-eight feet high, and cost £10,329, 16s. 6d. One of those strange memorials of Irish antiquity, a *cloghaun,* or pillar-stone, of considerable height, stands near it. The light, a catoptric fixed white light, visible for twelve miles, was first exhibited in 1841.

Until a light was set up on Tearaght, the *Skelligs Rocks* boasted of two lighthouses. The upper, at a height of three hundred and seventy-two feet above the sea, and forty-eight feet high, cost £25,721, 15s. 10d. The lower, forty-six feet high, one hundred and seventy-five feet above the sea, with a fixed white light, visible for eighteen miles, dates from 1826, and cost £22,000. Large portions of the rock had to be cut away to form a platform for sites of tower and dwellings. From the crosses and remains of ancient buildings still extant on the rock, there is reason to conclude that it was formerly the site of a small monastery dedicated to St. Michael. The Skelligs lie in lat. 51° 46′ 14″, at a distance of seven and a half miles from the shore. Miss Jean Ingelow has written a popular romance, with the title "Off the Skelligs," which contains some graphic sketches of the Kerry coast, and of these famous rocks.   For example :—

"The Great Skellig!   I had seen a picture of a rock —a hard material thing; I had read descriptions of its geological strata; I knew it was a thousand feet high; —but was *this* the Great Skellig?   I stood amazed. There was a pale glassy sea, an empty sky, and right ahead of us, in the desert waters, floated and seemed to swim a towering spire of a faint rosy hue, and looking as if, though it was a mile off, its sharp pinnacle shot up into the very sky.

"The 'westernmost point of British land, and out of sight of the coast,' was this that cruel rock on which the raging waves had driven such countless wrecks, and pounded them to pieces on its slippery sides?

"A boat was lowered.   Tom was going to row round it, though he said that, calm as the water was, it was still not quite safe to land.   To my delight he volunteered to take me with him; so I sent for my hat and cloak, and we rowed towards the great rock in the glorious afternoon sunshine.

"How often have I been disappointed in the outline of hills and mountains; they seldom appear steep enough to satisfy the expectation that fancy has raised.

"Here there was no disappointment.   The Great Skellig shot up perpendicularly from the sea—not an inch of shore; the clear water lapping round it was not soiled by the least bit of gravel or sand.   As we drew near, its hue changed; a delicate green down seemed to grow on it here and there. I sat in the boat and looked up, till at last its towering ledges hung almost over us, and its grand solitary head was lost, and the dark base showed itself in all its inaccessible bareness.

"Tom said to me, 'Do you see those peaks that look like little pinnacles?'

"I looked, and his finger directed me to a row of points about a third of the height of the rock, and projecting from it.

"'These points,' he continued, 'are as high as Salisbury Spire: when there is a storm, the wave breaks high enough to cover them with spray.'

"So sweet and calm they looked, serene and happy, I could hardly believe what I heard, nor picture to my heart the cries and wailing of human voices, the rending, pounding, and wrecking of human work that had been done on them, tossing from peak to peak, and ground on the pitiless rock, since first men sailed."

The *Calf Rock* is situated on the north-west coast of Bantry Bay, in lat. 51° 34′ 10″, and long. 10° 14′ 50″. Tradition relates that one of the sons of Milesius, leading an expedition to the invasion of Ireland, was wrecked with all his followers on this storm-beaten ocean fastness, which, in remembrance of the event, is to this day locally known as *Teach Dhoinn*, or "Don's House." Few spots on the Irish coast are more exposed than this to the violence of wind and wave.

In 1860–66 a lighthouse tower was erected on it, built of brickwork bound with iron casings, painted red with a broad white belt, and measuring one hundred and two feet in height from base to vane. It exhibited, at an elevation of one hundred and thirty-six feet above the sea, a holophotal light, revolving, with a flash every fifteen seconds. During a severe storm in January 1869 a large portion of the gallery was carried away. A new gallery of improved shape and less projection was then constructed, and the tower strengthened for thirty feet up with metal plates and an inner stratum of concrete. But a more furious gale assaulted the unfortunate lighthouse on the night of November 26, 1881, reducing it to a complete wreck, and endangering the lives of the keepers, who were detained prisoners on the rock for several days, no vessel being able to approach to their relief. The news of the calamity did not reach Dublin until late on the 28th. Within a few hours H.M.S. *Salamis* was despatched to render assistance, and was signalled by the lighthouse-men; but the sea was so rough that nothing could be done to rescue them, and in the afternoon of the 29th she returned to Bantry Bay. Next morning she made another effort to reach the rock, but in consequence of the continued violence of the weather, it was deemed unsafe to launch a boat. At the same time a cask of fresh water was floated to the keepers, who, it was feared, might be undergoing painful privations; and pending further attempts to take them off the rock, where they were exposed to the full fury of the storm, the rocket apparatus was brought into operation to supply them with food. The *Salamis* then returned to Buckhaven with the information that forty feet of the lighthouse had been carried away; but it was afterwards found that only the thirty feet remained which had been strengthened in 1867.

On the 3rd of December, the *Times* correspondent at Cork telegraphed that the six occupants of the wrecked lighthouse were still confined by wind and wave to their lonely prison. The principal lighthouse keeper, however, had contrived to send a communication ashore in an air-tight india-rubber bag, in which he gave full particulars of the disaster, and of the situation of himself and his colleagues. Two gunboats, the *Seahorse* and the *Amelia*, were cruising daily around the rock, though unable to get near it. The appliances on the rock which had formerly been used to enable the keepers to go ashore and return had shared the fate of the tower. They consisted of a gaff, sixty-three feet long, with a mesh attached to it.

The Calf Rock light and the Fastnet light—which also suffered severely during the gale—are the most important on the south coast of Ireland, being the chief guides to vessels arriving from America. With the exception, perhaps, of the Wolf, on the west coast of Cornwall, there is not on the shores of the United Kingdom a lighthouse which has to breast fiercer or heavier seas than those which strike on the Calf Rock. The tower was one hundred and fifty feet high, and Captain Boxer, inspector of Irish lighthouses, says that he has often seen waves break upon it which have gone into the air twenty feet higher. Strong currents from north and south here rush together, producing even in the calmest summer weather a ground-swell which makes it extremely dangerous to try to effect a landing. In 1868 seven men attempted it, and neither the boat they rowed in nor the men themselves were ever seen again. In shape the rock resembles an egg, the edges having been worn away by the waves until the whole surface has become quite conical. Its circuit at low water measures about one hundred yards.

The six keepers and assistants were rescued at length on the 8th of December by the gunboat *Seahorse*. The weather having moderated, she got near enough to throw to them a life-buoy with a line attached. This line having been made fast, another line was flung from the rock and picked up by a boat's crew which had approached the rock. The ropes were made secure; and at low water the men, one at a time, jumped into the sea, and were hauled on board the boat. They were soon transferred to the *Seahorse*, where stimulants and dry clothing were supplied. All the men were in a weak state, but having been conveyed ashore, and received proper attention, in a day or two they completely recovered.

The Calf Rock has since been abandoned as a lighthouse station, and a massive tower, with all the latest improvements, was erected, in 1889, on *Bull Rock,* lat. 52° 35′ 30″.* On *Roancarrig Island,* Buckhaven, eastern entrance of Bantry Bay, and on *Rock Island Point,* Crookhaven, north side of entrance, light towers were built in 1847 and 1867 respectively, the former by Mr. Halpin, the latter by Mr. Sloane. The former is sixty-two feet high, with a catoptric fixed white light; and the latter, forty-five feet high, showing a fixed light, with red and white sectors. Both are circular white towers of substantial construction. The Roancarrig tower is distinguished by a broad belt of red under the balcony.

LIGHTHOUSES IN THE CHANNEL ISLANDS.

The reader will not fail to remember that it was on the fatal rocks of the Casquets, off the coast of Jersey, that the ship *La Blanche Nef,* carrying homeward the son and heir of Henry I., Prince William, was wrecked in 1120. Prince William lost his life; and the royal father, it is said, never smiled again. Many good ships have since been destroyed upon this formidable reef; but since 1723 its position, to the great profit of the mariner, has been indicated by a warning light. The present building was reconstructed in 1877. It is built

---

* It is 49 feet high, 271 feet above high water, and shows a flashing light every fifteen seconds.

of stone, seventy-five feet high, and situated on the highest rock of the cluster, in lat. 49° 43′ 17″, and long. 2° 22′ 42″. The light shows three successive flashes of about two seconds' duration each, with about three seconds of darkness between each flash, and the third flash being followed by an eclipse of about eighteen seconds. A fog-siren is stationed here.

A noble tower of granite, one hundred and seventeen feet high, was erected in 1862 on the south-west point of the *Hanois Rocks*, Guernsey. It exhibits a red light, revolving in forty-five seconds.

Off the south-west extremity of the island of Jersey cluster the romantic rocks of the *Corbières*, flung out among the yeasty waters like the vertebræ of some gigantic ocean monster. They derive their name from the flocks of sea-cormorants (*corbière*) which breed among them. At low water they are accessible from the shore by a half-tide causeway, but at high water are entirely isolated. Their strange and varied outlines render them extremely picturesque. Tradition affirms that the *droit d'épaves*, or right of wreckage, whereby a stranded vessel became the property of the seigneur, was freely exercised of old, and ships lured by deceitful lights to destruction upon the Corbières. But their whereabouts is now indicated by a tall lighthouse tower of concrete, erected in 1874, in lat. 49° 10′ 40″. Its height is sixty-two feet, and the rock on which it stands is about sixty-five to seventy feet above high-water mark. It shows a fixed

THE CORBIÈRES LIGHTHOUSE.

white light, with two red sectors, visible nineteen miles; and in foggy weather a bell peals from its gallery thrice at intervals of half-a-minute. The light-house, and the two cottages for the accommodation of the keepers, was designed by Sir John Coode, C.E.

CHAPTER VII.

SOME FRENCH LIGHTHOUSES.

EVERY visitor to Havre will remember the two light-towers so prominently situated on the promontory known as the Cape de la Hève—lighthouses which have enjoyed the very rare distinction of being celebrated by a poet. For in one of his lyrics Casimir Delavigne apostrophizes them very prettily:—

> " Doux feux qui protégez et Thétis et la Seine,
> Sûrs et brillants rivaux des deux frères d'Hélène,
> Phares, je vous salue ; assurez à jamais
> Le commerce opulent de l'heureuse Neustrie ;
>   Fixez dans ma patrie
> L'abondance, les arts, tous les fruits de la paix."

The promontory itself has been embellished with a legend by Bernardin de St. Pierre, and is bold enough and romantic enough in its aspect to justify the fancies of legend and poetry. It rises abruptly from the wild waters of the Channel—"a precipitous mountain, composed of funereal strata of white and black stones"—the tomb of the fair nymph Héva, who died of sorrow for the loss of her mistress, the Seine, the daughter of

Bacchus and Ceres, who was metamorphosed into a flowing river to escape the pursuit of Neptune.

Cape de la Hève, the ancient *Caletes,* forms one of the boundaries and breakwaters of the estuary of the Seine. In the tenth century it projected much farther into the sea, and the banks of L'Eclat, now separated from it by a channel two thousand yards wide, was then a portion of it. The waters still continue their ravages, and it is estimated that they gain upon the land seven feet every year.

Tradition assigns to its two lighthouses an origin of great antiquity. As early as 1364, it is said, a tower, called the *Tour des Castillans,* was erected here, and a coal-fire was maintained upon its summit. This, however, fell into disuse; the tower was demolished, and the seamen sailing into the port of Havre were left without any assistance until, in 1774, the government of Louis XV., in compliance with the solicitation of the Chamber of Normandy, built the present edifices. At first coal-fire was the illuminant employed; but in 1781 each was fitted up with an apparatus of sixteen spherical reflectors, some focussing three and others two oil-wick burners, of which there were forty in all. The double paraboloidal reflectors of Bordier-Marceat, six to each lighthouse, were introduced in 1811 and 1814, their number being increased to ten in 1819. In 1845, the towers were restored and remodelled to fit them for receiving the improved dioptric apparatus in lanterns measuring twelve feet in diameter. A well-built row of dwellings was put up to connect the two towers. In 1866, the electric light was substituted for oil, the Alliance Company's "alternating current magneto-electric machines" being adopted for this purpose.

Both the towers are of equal elevation (seventy-five feet), and very handsomely constructed. From the balcony a view of great extent and impressiveness may be enjoyed. In brilliancy of colouring and variety of outline it has been compared to the pictures of sea and shore which the traveller commands at Corinth or Constantinople. When the air is clear and the sky cloudless, the spectator's eye ranges as far as Barfleur on the south-west; westward, to Honfleur, Trouville, and the picturesque little bathing-places on the Normandy coast; while far away in the distance lies Cape la Hague, with its memories of Admiral Russell's famous victory; and northward the gaze rests on the headland of Antifer, and the riven and sombre rocks of Etretat.

One of the most interesting of the French light-houses is that which illuminates the broad but danger-ous water-way between the Breton coast and the Roches-Douvres.

When M. Léonce Reynaud began his operations, he found himself confronted by difficulties almost as great as those which Smeaton conquered at the Eddystone and Stevenson at the Bell Rock. The rock selected for the site of the proposed lighthouse is one of a group which the sea covers at high tide. The men, therefore, could work only for a limited number of hours daily. **Again**: the currents in these waters were of great violence, their velocity **not** being less than eight knots **an hour**; and when their force is increased by the fury of a hurricane, the billows rage with a roar and **a** rush which fill the air like peals of thunder.

Huts for the masons were planted on **the** isle of Bréhat, about three leagues' distance from the rock. Here, **in** a sheltered little nook, a jetty of rough stones, about one hundred and seventy **feet** long, **was** constructed, to facilitate the transport of materials and the **landing** and embarking of **the men**. Quite a fleet was **engaged** in conveying **stores and** materials to the island. The granite came from **the** Ile-Grande, about ten leagues **to** the westward; the lime from the basin of the Loire; Saint-Malo furnished the timber; **and as** the wells of Bréhat were insufficient to supply the additional population, water, as **well** as provisions, **was** obtained **from** the mainland.

M. Reynaud's **"army of labour"** consisted **of** sixty **men.** During the **working** season these were lodged **on the** rock itself, or rather, at a short distance from it, on a platform of masonry, thirteen feet above high water, constructed between a couple of *aiguilles*, or needle-rocks, which the sea never covered. Several dwellings **were** erected **upon this** platform, together **with the timber** framework of **a** beacon, which exhibited a provisional light. The available area was very limited, and not an inch of **it** was wasted. In the beacon was arranged, besides the store-room and the light-keeper's lodging, a room for the accommodation of M. Reynaud. His bivouacking hut stood on the right. By blowing up a portion of the rock, it was found possible to erect a long but narrow barrack for the overseers. In front stood the kitchen and larder; at the side, the common dining-room; and in the rear, a sitting-room and bedroom for the men, in which the beds were set as close as possible, in two tiers. A third row was built up in the dining-room. Lastly, on a projecting crag to the left, was planted a small forge, which had but one defect, that it could not be kept lighted at high water.

At first the men were allowed full liberty in supplying themselves with provisions; but some cases of scurvy having broken out, M. Reynaud was compelled to impose upon them a carefully regulated bill of fare. He established, therefore, a canteen, binding down its lessee and manager to keep on hand not less than a six weeks' supply, as a precaution against bad weather, which might cut off all communication with the mainland. At this canteen each workman was required to obtain his rations. Due sanitary measures were also taken. The hammocks were exposed every morning in the open air, and once a week the living-rooms were lime-washed. Once a week, too, every man was compelled to take a sea-water bath. The result was that the terrible malady with an invasion of which the little colony had been threatened was stamped out, and the general health was in all respects satisfactory.

Every day, as soon as the tide had ebbed, work was begun, and the hours for meals were so arranged that no interruption took place while the tide lasted. When the rise of the waters rendered it necessary to desist, a bell gave the signal; and the men hastened, before they withdrew, to cover with a layer of cement, which hardened instantaneously, the masonry which had just been laid. Occasionally the tide flowed with unexpected rapidity, and then woe to the laggard and the unready! They were compelled to plunge into the rolling waves, and make their way as best they could to the place of shelter, where in their dripping condition they were always received by their fellows with shouts of laughter. This frequently-recurring scene was almost the only amusement of the little colony, and they made the most of it.

So far as the erection of the light-tower was concerned, the chief difficulty was necessarily experienced in laying the submarine foundations. As soon as they had reached high-water mark, the men not only carried on their labours with greater facility and convenience, but were relieved from much risk of danger. The rock on which they worked consisted of an exceedingly hard black porphyry, in which an excavation was effected of about twenty inches in depth and thirty-eight feet in diameter. In this excavation were laid the first courses of masonry. The lighthouse itself, one hun-

dred and fifty‑five feet in height, is divided into
two principal parts. The tower is built up solid to a
point three feet and a quarter above the level of the
highest tides, and forms a massive and impregnable
substructure, from which rises another elegant and
substantial tower, crowned by a spacious lantern. The
work occupied six years. The first was employed in
surveying the site, and in preparing and maturing
plans; the second, in the erection of the huts and
beacon and excavation of the rock; the third, in build‑
ing up the solid masonry; in the fourth, the light‑tower
was raised as far as the first gallery; in the fifth, it
was carried a little above the cornice; and, finally, in
1839, the lantern was completed and the light exhibited.
The total cost of this noble work was 577,984 francs.

Standing erect and solitary, in the midst of a wide
waste of waters, the lighthouse of the Héaux of Bréhat
produces a strong impression of austere magnificence.
Michelet says of it that it has the sublime simplicity
of a gigantic sea‑plant. Massive, immovable, silent, it
seems, from one point of view, a defiance flung by the
genius of man to the fierce demon of the storm. The
best description of it with which we are acquainted is
that so graphically drawn by M. de Quatrefages, in his
"Rambles of a Naturalist." He visited it from Bréhat
in a boat manned by six sturdy seamen, on a fine
October day, and was evidently stirred to a high pitch
of enthusiasm. Here is his account of what he saw
and felt:—

"The nearer we approached to Héaux," he says, "the taller seemed its beacon-tower, which stood forth from its substructure, with lofty granite column and glass lantern, protected by that magical rod which is able to attract and safely guide to earth the destructive force of the thunderbolt. We landed, and at once began an examination of this colossal bulk raised by the hand of man on the Epées de Tréguier, which, formerly the terror of the seaman, has now become his protecting guide through the darkness and the storm.

"The Héaux Lighthouse would be looked upon as a remarkable monument even in our principal towns; but standing as it does alone in the midst of ocean, it acquires from its very isolation a character of grand severity which powerfully influences the mind. Figure to yourself a wall of granite, where the winds and waves forbid even the hardiest ferns to take root, with here and there a twisted and deeply-worn crag projecting beyond the rest of the mass. Here it is that the architect has laid the foundation of the tower. The base, which is conical in form, is surmounted by a circular gallery. The lower portion curves gracefully outwards, and extends over the ground like the root of some vast marine plant springing up from the foundation stones, which have been deeply imbedded in the rock. Upon this base or substructure rises a column twenty-six feet in diameter, which is surmounted by a second gallery, with supports and stone balustrades that remind you of the portcullis and battlements of some

medieval keep. From base to summit this column or
tower is composed of large blocks of a whitish granite,
arranged in regular courses, and carefully dovetailed
into one another. For a third of the total height of
the building the rows of stones are bound together by
granite joggles, which at the same time penetrate into
the two superposed stones. The stones have been cut
and arranged with such precision that hardly any
necessity has arisen for the use of cement, and it has
been employed only in filling up a few imperceptible
voids. Hence, from base to summit, the lighthouse
forms one solid block, which is more homogeneous and
probably more compact than the rocks which support
it. The platform that crowns this splendid column, at
an elevation of upwards of one hundred and forty feet
above high-water mark, is surmounted by a stone
cupola, at once solid and graceful, resting upon pillars
which are separated by large panes of glass. It is
within this crystal frame that the beacon is lighted,
which may distinctly be seen from every direction at a
distance of twenty-seven miles.

"At low tide the sea leaves a space of several hundred square yards uncovered round the base of the
edifice; at high tide, it entirely surrounds it. Then
it is that the tower of Héaux rises in its solemn soli-
tariness from the bosom of the waves, like a standard
of defiance unfurled by man against the spirit of the
storm. One might almost fancy at times that the
heavens and the sea, conscious of the outrage offered to

them, were leagued together against the enemy which seems to brave them by its imperturbability. The north-west wind roars round the tower, darkening its thick glass windows with torrents of rain and drifts of snow and hail. These impetuous blasts bear along with them gigantic billows, which not unfrequently with their foamy crests reach the first gallery, but slide away from the round polished surface of the granite that presents no points of adhesion, and hurling their spray even above the cupola, plunge forward with thunderous roar against the rocks of Stallio-Bras or the boulders of Sillon. The tower supports these tremendous attacks without injury, though it bends, as if in homage, before the might of its foes. I was assured by the keepers," says M. de Quatrefages, "that, during a violent storm, the oil in the lamps of the highest rooms shows a variation of level exceeding an inch, which would lead us to assume that the summit of the tower describes an arc of about a yard in extent. This very flexibility is, however, in itself a proof of durability. At all events, we meet with similar conditions in several monuments which have endured for ages the inclemency of recurring winters. The spire of Strasburg Cathedral, in particular, bends its long ogives and slender pinnacles beneath the force of the winds, while the cross on its summit oscillates at an elevation of more than four hundred and fifty feet above the ground.

"To construct a monument upon these rocks, which are apparently the centre of all the storms which rage

on that part of our coasts, was like building an edifice
in the open sea. Such a project at first sight seemed
almost impracticable. After their third season of labour,
the workmen completed the foundations of the **tower**
and fixed the keystone of the cupola. In vain did
difficulties of every kind combine with the winds and
waters to oppose their work; human industry emerged
victorious from the struggle, **and no serious accident**
disturbed the toilers in the joy **of** their triumph. **On**
one occasion alone was science at fault. In order to
facilitate the arrival of **the** stones, which **had to be**
brought from a distance of several leagues and cut at
Bréhat, the skilful engineer who had furnished all the
plans and superintended their execution resolved **to**
construct a wooden pier **for** the landing of the stones
where they were required. Some veteran seamen ob-
jected to the project as impracticable; but **M.** Reynaud,
unfamiliar with the sea, and proud, moreover, of having
stemmed the current of rapid **rivers,** relied on the
stability of the massive pillars which he had clamped
together with bronze and iron. But he was soon compelled to acknowledge his mistake. The first storm
sufficed to scatter over the waters all his ponderous
and solid materials, like so many pieces of straw. A
crane was therefore raised upon the summit of a rock,
to which boats **could be** moored; and the materials
were **then** hauled **up to a** railway which had been
thrown over **a** precipice that separated this natural
landing-place from the site **of the tower.**

"Now that we have admired the lighthouse externally, follow me into the interior by the help of these steps, which have been formed by the insertion of bars of gun-metal into the stone. Pause for a moment to admire the ponderous bronze doors which hermetically seal the entrance, before we pass into those vaults which look as if they had been cut out of the solid rock. On the first story we are surrounded by stores of wood and ropes and workmen's tools. On the next we see cases of zinc, which, we are told, contain oil to feed the lamps and water for the use of the keepers. On the third story, with the first or lower gallery, is the kitchen with its pantry and larder. Of the three apartments appropriated as their bedrooms and sitting-rooms, we have nothing to say, except that they are very simple and clean. On the seventh story we rest for a few moments in the little octagonal saloon set apart for the engineers on their visits of inspection. Here, in the midst of the ocean, you will find the comfort and almost the elegance of a Parisian apartment.

"Let us now regain the spiral staircase which has brought us to this elevation, and it will carry us still higher to that portion of the building which is more particularly designed to fulfil its special object. The eighth story is filled with vessels of oil, sheets of glass, revolving lamps, some excellent meteorological instruments, a thermometer, a barometer, and a chronometer. Here the spiral staircase terminates in a flattened arch,

which supports a slender pillar cut into steps—the
only means of access to the watch-room above, where
the men take it by turns to keep watch every night.
You will be surprised on looking round to observe that
this room is lined with different coloured marbles—the
roof, the walls, and even the floor. But this luxury,
which may appear to you so much out of place, has
been introduced from necessity. The lighting apparatus enters the apartment through a circular aperture in
the ceiling; and hence the most absolute cleanliness,
which can be secured only by means of perfectly
polished surfaces, is indispensable."

From Bréhat we now transport the reader to the
little seaport of Sables-d'Olonne, and proceed on a
visit to the lighthouse, erected in 1861, on the dangerous rock of the *Grande Barge d'Olonne*, situated about
a mile and a furlong from the shore, in the midst of
currents and counter-currents of extreme violence. Its
foundation is almost completely submerged, and during
high tides the waves mount to a height of one hundred
feet.

Building operations were begun in 1857, and completed in 1861; but in these five years, owing to the
difficult conditions under which they were prosecuted,
only one thousand nine hundred and sixty hours could
be devoted to continuous labour. Yet, so well understood are the principles on which these structures are
based, and so infinite are the resources of engineering

science, that this comparatively brief period proved sufficient.

The entire cost of the building is stated at 450,000 francs, equal to £18,000. It was erected under the supervision of M. Reynaud, inspector-general, and M. Forestier, engineer-in-chief, the material employed being granite; diameter at the base, 39·37 feet, tapering with a curved outline to 21·23 feet beneath the balcony. The door-sill is thirteen and a half feet above high-water mark of the highest tides; and up to this point the tower, with the exception of receptacles for coals and fresh water, is solid. Above, the main shaft or tower, with an interior diameter of 11·48 feet, is divided into five stories by vaults of brick. It has a substantial cornice and parapet of granite; and from its topmost platform rises the turret, 6·56 feet high and 8·2 feet in diameter, which supports the lantern. The catadioptric illuminating apparatus which is in use here produces a white light, varied by red flashes every three minutes.

Opposite the rising watering-place of Arcachon projects the picturesque promontory of *Cap Ferret*, forming one side of the entrance to the Arcachon basin, in lat. 44° 39'. On its summit stands the tall tower of a lighthouse of the first class—a circular edifice rising from a circular base (which contains the keepers' room) to an elevation of one hundred and fifty-six feet. It is not divided into stories, but a spiral staircase winds up the interior to the lantern-room. The bars of the lantern, we may note, are set horizontally and vertically, instead of laterally and spirally as in England, or diagonally as in Scotland; and in this respect it is inferior, for horizontal bars cast shadows and impair the light. A strong wire network surrounds and encloses it, as a protection from the sea-birds, which are attracted in great numbers. As many as two hundred have been killed in a single year by dashing against the panes.

This light has three keepers. No provision had been made for lodging their families, but some thirty years ago they obtained permission to build huts for themselves and their families. No means of communication exist between the lantern-room and the keepers' room. The man on duty is allowed an arm-chair, but no books. It has been remarked that the reverse obtains in England, where books are provided, but no arm-chair.

The lighthouse at *Biarritz*, within the shadow of the pine-clad Pyrenees, is well worth a visit, if only for its romantic position, and the beauty of the views both of land and sea which it commands. It is built on the rocky headland of St. Martin's Point, two and a half miles south-west of the mouth of the river Adour; and standing on the brink of the cliff, which is about a hundred feet high, shows in a very handsome and stately manner. Inside, as at Cap Ferret, the ascent to the lantern is by a spiral staircase. The floor is of coloured marbles. The illuminating apparatus, dioptric, first-order, shows a revolving light, white and red, revolving in twenty seconds, and visible for twenty-two miles.

On the isle of Ré, so sadly famous in English history as the scene of the failure of the Duke of Buckingham's expedition in the reign of Charles I., is situated the *Tour des Baleines,* in lat. 46° 25'. It is built of yellow stone, with ornaments of dark-gray granite. The tower is octagonal in shape, and rises from a block of buildings, two stories high, which is divided into numerous apartments. These are well fitted and furnished. The oil storeroom and the men's workroom are quite remarkable for the elegance of their fittings, their pavements of coloured marbles, tables of the same material, glass cases for the necessary tools, spare lamps, and the like. The tower measures one hundred and sixty-four feet from base to vane, and its focal plane is one hundred and sixty-six feet above the sea. It is hollow, and ascended by a spiral stair. Below the lantern is a room, with a spare lamp ready for use, and a bed for a keeper. This room is wainscotted with flowered oak, and exceedingly handsome. Coloured marbles are employed as a flooring for the lantern, and line its walls to the height of six feet. The gallery outside is broad, and protected by a solid rail. Total cost of the Tour des Baleines, £14,010. An electric light apparatus was introduced in 1882, and it now shows a succession of groups of four flashes, with an interval of ten seconds between two groups.

Relative to the lighthouse service in France, we may state a few particulars. Its centralization dates from

the time of the National Convention. Previously, the
existing lighthouses had belonged to local corporations;
but a law of the 15th February 1792 placed them
under the control of the Ministry of the Marine, but
intrusted the execution of the works to the Minister of
the Interior. An imperial decree of March 7, 1806,
transferred the service entirely to the Ministry of the
Interior, and afterwards to the Conseil Général des
Ponts et Chaussées; but it prescribed a concerted
action between the French Admiralty and Home Office
for the organization of establishments, and out of this
grew eventually the Commission des Phares, or Lighthouse Commission, which consists of three naval officers,
three members of the Institute, three general inspectors
of roads and bridges, and the Minister of Public Works
as president. This Commission is called on to decide
(subject to the approval of the minister) on all questions relative to the establishment of lighthouses,
buoys, and beacons, and their maintenance and improvement. It administers annually an expenditure of
about £50,000.

The French system allots three keepers to lighthouses of the first order (the lights of which have a
range from twenty to twenty-seven miles), two to those
of the second and third order (six to twenty miles),
while lights of the fourth order, or fanaux, have but
one keeper each. In rock lighthouses there are always
three keepers, to whatever order they may belong, so
that the service may never be exposed to interruption.

and the keepers may have regular periods of rest.
Lighthouses of the first order, in isolated positions,
are provided with four keepers.

There are seven classes of keepers, whose wages are
as follow :—

Maîtres de Phare	....	£40
Gardiens de 1re Classe	....	34
,,	2me ,,	 31
,,	3me ,,	 28
,,	4me ,,	 25
,,	5me ,,	 22
,,	6me ,,	 19

They are supplied also with lodging, fire, and light.
For good conduct they obtain rewards, of which the
maximum is fixed at a month's allowance. They are
entitled to a retiring pension, towards which five per
cent. is deducted from their wages.

We subjoin a list of the more remarkable of the
French lighthouses :—

Cape Grisnez, lat. 50° 52′, long. 1° 35′; electric light,
showing three white flashes at three-second intervals,
followed by an interval of about twelve seconds, in the
middle of which is seen a *red* flash ; circular tower, 79
feet high ; first lighted in 1837. To the French lighthouse service Cape Grisnez bears the same relation as
the South Foreland to the English service ; important
experiments are constantly being performed there. Its
light is visible from the English coast.

Cape La Hève, lat. 49° 31′; two lighthouses, each 66

feet high; **each** square built, of stone, with **focal plane** of lantern **397 feet** above the sea; range of light, **27 miles.** The electric light is in use here.

Fatouville, lat. 49° 25′; **fixed** white light, with red flash every three minutes; visible **22 miles**; octagonal stone tower, **105 feet high, erected on the cliffs upwards of 300 feet above the sea**; dates from 1850.

Cape Léve, between Barfleur and Honfleur, lat. 49° **42′; fixed white light, with red** flash every three **minutes, visible 13 miles**; square tower, **105 feet** high; **erected in 1858.**

Cape la Hague, lat. 49° 43′. Off **this** headland Admiral Russell won his great victory over the French fleet under **M. de** Tourville in 1692. The lighthouse, a circular tower **of** stone, **154** feet high, **was** erected in **1837 on the** summit of the Gros du Raz rock, half **a mile from the cape; shows a fixed** white light, visible **18 miles.**

Grand Lejon, **on the** rock, lat. 48° **45′,** long. 2° 40′; white **and** red, alternately fixed and flashing—the fixed white light visible 12 miles, and the fixed red light **8 miles;** the flashing white 17 miles, and the flashing red **12 miles. The** tower **is** 76 feet high; erected in 1881.

Héaux de Bréhat, **on the north-east side of** Les Héaux ledge, lat. 48° **55′; fixed** white and red light, visible 18 miles; circular **tower of** granite, 159 feet from base to **top;** lighted **in 1839;** cost of construction £23,219.

Ile de Bas, lat. 48° 45′; **white** light, revolving every minute; visible 21 miles; round tower, 131 feet high; erected in 1836.

Ile de Ushant, lat. 48° 28′, on Creac'h Point, north-west side of island; white and red light, revolving; visible 22 miles; circular tower of stone, painted with horizontal belts of black and white, 151 feet high; erected in 1863.

Ardmon Rock, lat. 48° 3′; fixed white light, visible 16 miles; tower, 110 feet high; first lighted in 1881. In foggy weather a trumpet gives a blast at minute intervals of about five seconds' duration.

Penmarch, on the point, near Church of St. Pierre, lat. 47° 48′; white light, revolving in thirty seconds; visible 18 miles; circular tower of stone, 131 feet high; erected in 1835.

Belle Ile, Goulfar Bay, lat. 47° 19′; white light, revolving every minute; circular tower of stone, 151 feet high; erected 1857.

La Tour, lat. 47° 18′; white light, revolving; visible 15 miles; circular tower, 92 feet; erected 1846.

Barge d'Olonne, on Grand Bank, lat. 46° 30′; fixed white light, with red flash every three minutes; visible 14 miles; circular tower, 84 feet; lighted 1861.

Ile d'Olèron, Chassiron, on north-west point, lat. 46° 3′; fixed white light, visible 19 miles; circular tower, 141 feet; erected in 1836.

Gironde River, Coubre Point, lat. 45° 42′; fixed white light, visible 17 miles; tower painted dark brown, 98 feet high; 1860.

Gironde River, Terre Nègre, lat. 45° 39'; fixed white light, visible 15 miles; circular black and white tower, 75 feet; 1860.

Gironde River, St. Pierre, N.E. by E. from Chay Lighthouse, lat. 45° 37'; fixed red light; square tower, painted with belts of red and white, 115 feet high; lighted in 1873.

Gironde River, Tour de Cordouan, on rock, lat. 45° 35'; revolving light, red and white, every minute; circular tower, 207 feet high; erected 1854.

Cap Ferret, lat. 44° 39'; fixed white light, visible 19 miles; circular tower of granite, 156 feet; lighted 1840.

Contis, lat. 44° 6'; white light, revolving every half minute; visible 19 miles; circular tower, 125 feet high; erected 1863.

Biarritz, St. Martin's Point, lat. 43° 30'; white and red light, revolving in twenty seconds; visible 22 miles; circular tower of brick, 144 feet high; erected 1861.

CHAPTER VIII.

LIGHTHOUSES OF THE UNITED STATES.

IN the United States the lighthouse service is under the charge of a Board legally organized on the 31st of August 1852. This Board comprises the Secretary of the Treasury, who is *ex officio* president, two officers of the navy of high rank, two officers of engineers of the army, two civilians distinguished by their scientific attainments, and one officer of the navy and one officer of engineers of the army as secretaries. The clerical work of the board is executed by five clerks.

No officer of the army or navy serving on lighthouse duty receives any other compensation than that to which he is entitled in his grade in the service.

The ocean and lake coasts are divided into twelve districts, each of which is under the charge of an inspector, who is an officer of the army or navy, and reports directly to the Board. In each district are also stationed superintendents of lights, who are collectors of the customs, and whose duties are, to disburse the money for the ordinary expenses of the establishment, on the certificates of the inspectors; to nominate keepers to the Secretary of the Treasury; and, in the sickness or necessary absence of the inspector, to perform his duties.

The inspectors and superintendents are directly responsible to the Board, but are in no respects responsible to each other.

From time to time engineers are selected from one or other of the corps of engineers of the army, who are either inspectors of lights, and incidentally engineers of lighthouse construction in their respective districts, or do duty as lighthouse engineers exclusively. Under their superintendence all lighthouse structures are built, after the designs have been submitted to and approved by the Board. Funds for the lighthouse service are voted by Congress, and no tax or toll is levied for its support. The annual cost would seem to be about £140,000, with about £20,000 for buoys and beacons.

In selecting sites for lighthouses, the Board is led to its conclusions by careful consideration of (1) the position of the lighthouses with regard to the necessities of navigation; (2) the possibility of obtaining a proper foundation. Similar principles are observed in deciding upon the stations of floating lights. The classification adopted is simply this: primary sea-coast lighthouse stations; secondary sea-coast and lake-coast lighthouse stations; and sound, bay, river, and harbour lighthouse stations. To first-class lights the staff allotted is—one keeper at £120 a year, and two assistants at £72 each; second-class lights—one keeper at £100, and one assistant at £60; third-class—one keeper at £90, and one assistant at £60; fourth, fifth, and sixth classes—one keeper at £80. There are no allowances.

The optical system in use is the lenticular; and this is preferred, because (1) the primary sea-coast lights are more powerful than reflector lights can be made; (2) for the same quantity of light the expenditure of oil is less; and (3) the annual cost of repairs and labour in the lens system is less than it is in the reflector system. Oil-lights are almost exclusively used; but electricity has been employed in the case of the Statue of Liberty at New York. The sole oil employed is sperm.

An American lighthouse of the first class is estimated to cost (including £1,600 for the illuminating apparatus) £8,600. A light-vessel will cost, when complete for service, about £4,200; and its crew will consist of a keeper at £200 per annum, first mate £96, second mate £72, and eight seamen at £43, 4s. each, with one ration for each person. The annual maintenance of a light-ship is put at £1,300 (including £250 for repairs).

The American lighthouses, having been constructed on the English model, do not call for detailed description, but we subjoin a few particulars of some of the most important:—

On the Atlantic sea-board the most northerly is the *St. Croix River*, situated on Dochet Island, lat.

45° 8′, erected in 1856; and the southernmost is the *Brazos Santiago*, on Isabel Point, lat. 26° 5′, erected in 1852,—eleven hundred and forty-three geographical miles apart. The total number of lights along this extent of coast is four hundred and seventy-four (fixed and floating lights); but, including those of lakes, rivers, etc., the aggregate rises to sixteen hundred and twenty-eight (of which, however, only fifty-one belong to the first class). This is exclusive of twenty-three lightships. Floating lights are not much in favour in the United States, and, wherever possible, permanent lighthouses have been substituted for them.

Petit Manan Island, lat. 44° 22′, off coast of Maine; white light, with flash every two minutes, visible 17 miles; gray tower 109 feet high; first lighted in 1817. *Every lighthouse on the coast of Maine is provided with a fog-signal.*

Seguin, on the island off Kennebec River, lat. 43° 42′; fixed white light, visible 22 miles; gray tower, 35 feet high; first lighted in 1795, and therefore one of the oldest of the American lighthouses.

Bom Island, off York Harbour, lat. 43° 7′; fixed white light, 18 miles; gray tower, 123 feet; erected in 1812.

Portsmouth, inner entrance of harbour, south-west side; fixed white light; 60 feet high; first lighted in 1789.

Cape Ann, Thatcher Island, Massachusetts, in lat. 42° 28′; fixed white light; visible 19 miles; two towers, each 102 feet high; and each first lighted in 1790. The two towers are 295 yards apart.

Boston Bay, Little Brewster Island, north side of main outer entrance of harbour, lat. 42° 40'; white light, revolving every half minute, visible 17 miles; circular tower, painted white, with black lantern; first lighted in 1716.

Plymouth, Garnet Point, lat. 42°; fixed white light, visible 16 miles; octagonal tower, 35 feet; dates from 1769.

Cape Cod, Massachusetts, lat. 42° 2'; fixed white light, visible 20 miles; white tower, lantern painted black; 53 feet from base to vane; centre of lantern 195 feet above the sea; erected in 1797.

Nantucket (Great Point), lat. 41° 25'; fixed white light, visible 14 miles; white tower, with black lantern; 60 feet high; erected in 1784.

Holmes Hole, Vineyard Sound, lat. 41° 31'; fixed white light, visible 14 miles; white tower, black lantern; erected 1827.

Whale Rock, Rhode Island, lat. 41° 27'; fixed red light, visible 14 miles; circular tower, 75 feet; erected 1882.

Montauk Point, Long Island, lat. 41° 4'; white light, with flash every two minutes, visible 19 miles; white stone tower, 97 feet high; erected in 1795.

Great West Bay, Pondquogue Point, New York, lat. 40° 51'; fixed white light, visible 19 miles; red tower, 150 feet high; erected 1857.

Fire Island, south side of Long Island, lat 40° 38';

white light, revolving every minute, visible 19 **miles; yellow** tower, **152** feet high; erected in 1826.

Sandy Hook, south side **of** entrance **to** New York Bay, **lat. 40° 28′,** long. 74°; **fixed** white light, visible 15 miles; white tower, **77 feet high;** erected in 1764.

Barnegat, on north end of Long Beach, lat. 39° 46′; **white** flashing **light** every **ten** seconds, visible 19 miles; **upper** half painted red, **lower** half painted **white;** 150 feet; erected 1834.

Absecon, **south side of** inlet, lat. 39° 22′; **fixed** white light, visible **19** miles; white tower, painted with horizontal stripes of red; 150 feet high; erected in 1856.

Cape May, Delaware Bay, lat. 38° 56′; white light, **revolving every 30** seconds, visible 19 **miles;** gray **tower,** with red lantern; 159 feet high; erected in 1823.

Cape Henlopen, Delaware Bay, lat. 38° 47′; fixed **white light, with** red sector, visible 17 miles; white tower, **with** lantern painted black ; **82** feet; erected in 1789.

Port Penn, **Delaware Bay,** lat. 39° 30′; fixed white light; building painted black ; 120 feet high ; erected **in 1877.**

Assateague Island, New Jersey, lat. **37°** 55′; fixed white light, visible 18 miles; **red tower,** 129 feet; erected 1833.

Chesapeake Bay, Virginia. The numerous shoals in this bay are indicated by screw-pile lighthouses **(on** Mitchell's system), and lights on ordinary piles,—about **29 in all.**

Currituck Beach, North Carolina, lat. 36° 23′; white light, with red flash every minute and a half, visible 19 miles; red tower, 150 feet high; erected in 1875. This lighthouse is drearily situated among high and glittering sand-hills, on a coast destitute of vegetation, and utterly unpicturesque.

Booby Island, north of Cape Hatteras, lat. 35° 49′; fixed white light, visible 19 miles; tower 150 feet high, painted with horizontal bands of black and white; erected in 1872.

Cape Hatteras (about two miles from south extremity), lat. 35° 15′; white light, flashing every ten seconds, and visible 20 miles; the tower is 189 feet from base to focal plane of lantern, and is painted with black and white spiral bands, but uniform red from base to a height of 27 feet; erected in 1798, rebuilt in 1870.

Cape Look-Out, lat. 34° 37′; fixed white light; tower 150 feet high, chequered in black and white; erected in 1812.

Cape Fear, lat. 33° 52′; red light, flashes every half minute, visible 16 miles; pyramidal white building, 96 feet high; erected in 1818.

Cape Romain, on Racoon Cay or Key, South Carolina, lat. 33° 1′; white light, revolving every minute, visible 18 miles; red tower, 150 feet high; erected in 1866. A light was first exhibited here in 1827, and the old tower is still standing, painted white, and 65 feet high.

Morris Island (for Charleston), lat. 32° 42′; fixed

white light; tower, with black and white bands, 150 feet; first erected in 1767, rebuilt in 1876.

Hunting Island, south side of entrance to St. Helena Sound, lat. 32° 23'; white light, revolving every half minute, visible 19 miles; tower 121 feet high, painted black in the upper part and white in the lower; erected in 1767, rebuilt in 1876.

Tybee Island, south side of entrance to Savannah River, lat. 32° 1'; fixed white light; octagonal white building, 134 feet high; erected in 1793, restored in 1867.

St. Simon Island, lat. 31° 8'; white light, with red and white flashes every minute; white tower, with black lantern; 100 feet high; erected in 1811, rebuilt in 1872.

St. Augustine, lat. 29° 53', north end of Anastasia Island; white light, with flash every three minutes; conical tower, 150 feet high, painted with black and white spiral bands; erected in 1823, rebuilt in 1874.

Cape Canaveral, lat. 28° 28'; white light, revolving every minute; tower 134 feet high, embellished with horizontal bands of black and white; erected in 1847. Along this part of the coast, from St. Augustine to Saint Cay, there are nine or ten primary lighthouses, all fitted with dioptric apparatus of the first order, namely — Cape Canaveral, Jupiter Inlet (94 feet), Fowey Rocks (115 feet), Carysfort Reef (112 feet), Alligator Reef (149 feet), Sombrero Cay (149 feet), Ammion Shoal (115 feet), and Sand Cay (121 feet).

Dry Tortugas, on the Loggerhead, or South-west Cay, lat. 24° 38′; fixed white light, visible 18 miles; tower 150 feet high, the lower part painted white and the upper black; erected in 1858.

Pensacola Harbour, near Fort Barrancas, on the north side of Pensacola Bay, lat. 30° 21′; white light, revolving every minute, visible 21 miles; tower 160 feet high, erected in 1858, the upper two-thirds painted black, the lower third white.

Sand Island, off the Alabama coast, about five miles south-south-west of Mobile Point, lat. 30° 11′; fixed white light, visible 12 miles; conical tower, painted black, 125 feet high; erected in 1873.

Mississippi River (mouth of), *South Pass* (west side of), lat. 29° 1′; white light, flashing every five seconds, visible 16 miles; tower, painted red, 105 feet high; rebuilt in 1881.

Mississippi River, on a low marshy island, on the west side of the pass, in lat. 28° 58′; fixed white light, visible 17 miles; skeleton tower, painted black, 126 feet high; reconstructed in 1873. This and the preceding lighthouse are the only two lights of the first class (dioptric, first order) on the Alabama, Mississippi, Louisiana, and Texan coasts, from lat. 30° 11′ to lat. 36° 5′.

Timbalier, near east end of Grande-Terre Island, lat. 29° 1′; white light, with red flash every minute, visible 16 miles; black tower 118 feet high; erected 1856, rebuilt 1874.

Sabine **Pass**, Brant Point, **east side of entrance to** Calcasieu River, lat. 29° **43′; white** light, **with flash every minute** and a half, visible **15 miles; octagonal white tower, 75** feet high**; erected** in **1856.**

Bolivar Point, **north side of** entrance **to Galveston,** lat. 29° 22′**; fixed white light, visible 17 miles; tower 110** feet high, **painted with horizontal bands of black and** white**; erected in 1872.**

CHAPTER IX.

LIGHTHOUSES IN OUR COLONIES AND DEPENDENCIES.

TO survey the extensive coast-line of our various colonies and dependencies,—Canada, the West Indies, the East Indies, Australia, New Zealand, Tasmania, with all their outlying possessions, besides the numerous islands sprinkled over ocean's broad breast which acknowledge the British flag,—and to indicate the fixed and floating lights which minister to the safety of navigation and the development of commercial enterprise all along this vast range of sea, would occupy, however briefly and imperfectly undertaken, a much more considerable space than we have here at our disposal ; and we must be content with particularizing only a few of the larger and more important.

Far away in the North Sea lies the rock-island of *Heligoland*, which we captured from the Danes on the 5th of September 1807. Here, on the highest summit, a light has been exhibited from a time unknown,—an oil light since February 1811,—the tower, which is of stone, circular in shape, and strongly built, having been restored and remodelled by the British Government, at a cost of £8,618, 14s. 7d. From base to vane it measures seventy-five feet, and its focal plane rises two hundred and fifty-eight feet above the sea. It was designed by Mr. David Alexander, and shows a fixed white light, visible for sixteen miles.

We now bend our course in a very different direction, traversing the English Channel, crossing the Bay of Biscay, passing Cadiz and its busy port, and, with a throb of patriotic pride, pausing for a moment to contemplate "the vision of the guarded mount," the thrice-famous *Rock of Gibraltar*—the first of the great strategic ports which guard our highway to India. The Bay of Gibraltar lies between two picturesque headlands—Cabrita in Spain and Europa Point on the rock. At the seaward extremity of the point a lighthouse, built by the Royal Engineers—a massive tower of limestone, eighty-one feet high—was first lighted on the 1st of August 1851. Its illuminating apparatus is first-order dioptric, and shows a fixed white light.

COLONIAL LIGHTS are generally under the jurisdiction of the respective colonies, though in some few instances they have been erected and are maintained by imperial funds. Among those which have been so erected, and are now or were until lately so maintained, wholly or in part, are the following :—

Bahamas—Gem Cay, Cay Sal, Abaco, Great Isaacs, Cap Lobos, Bird Rock, Castle Island, Turk Island, Inagua Great Island.

Cape of Good Hope—Roman Rocks, South Point.
Ceylon—Great Basses.
Newfoundland—Cape Race.
Falkland Islands—Cape Pembroke.
Western Australia—King George's Sound.

A few notes upon these may prove of interest to the reader.

The Bahamas are all low, consisting of coral-rock covered with a thin layer of soil. On the Atlantic side they rise sheer from the ocean depths, but in the opposite direction form vast submarine reefs, involving great danger and difficulty to navigation. They are therefore well lighted.

On *Gem Cay*, lat. 25° 34', is situated a conical tower of stone, 70 feet high, the upper part of which is painted red, the lower white; erected in 1836 at a cost of £3,077. Shows a white light, revolving every minute and a half.

In the same year was lighted up the station at *Abaco*—a conical stone tower, painted red and white like that at Gem Cay, measuring 85 feet from base to vane; cost £3,967; white light, revolving once a minute. A circular tower, with red and white bands, 77 feet high, was erected on *Little Guana Cay* in 1863.

In 1839 a conical limestone tower, 58 feet 9 inches high, painted red and white, was erected on *Cay Sal* at a cost of £3,804. Fixed white light, visible for about 15 miles.

Mr. Alexander Gordon (in December 1860) pub-

lished some curious facts respecting the irregularities
of the Bahama lights, or light-keepers. He showed
that fourteen times, in a few months, the light at Little
Guana Cay had been "out." Bad ventilation and the
wind blowing down had extinguished it, much to
the advantage of those who lived by wrecking—that
is, the greater part of the population of the Bahamas.
It was found upon inquiry that "the cowl" had stuck
fast, and the keepers would not liberate it. One
night, in 1856 or 1857, this lighthouse was in total
darkness until about eleven o'clock. The reason
assigned by the keepers was the danger of passing
from their dwellings to the tower in a storm. During
that interval of darkness, and because of it, a vessel
was wrecked on a cay four or five miles off, and all
hands perished. On the following morning the light-
keepers were busily engaged "enriching themselves
from the wreck."

Great Isaacs Island, another of the Bahamas, was
first lighted on the 1st of August 1859. This is one
of Mr. Alexander Gordon's iron lighthouses. It was
constructed in England, sent out in parts, and put
together under the supervision of a resident English
engineer. Total cost, about £14,300. It is painted
with broad belts of red and white; from base to vane
measures 145 feet; and exhibits a brilliant light, revolving every half minute, the range of which is 16
miles. Previous to the erection of this light, twenty-
five ships were cast away on this island in twelve

years, and in the same period nearly five hundred on various islands of the Bahamas group.

The iron lighthouse on *Lobos Cay*, lat. **22° 22′**, like that on Great Isaacs, was designed by Mr. Alexander Gordon, constructed in England, and erected **under** the superintendence of a resident engineer sent out from this country in 1860. It is 150 feet in height, painted with broad belts **of red** and white, and cost about £19,000, exclusive of the lantern and illuminating apparatus, which cost £1,248, 8s.

On *Bird Rock*, in Crooked Island Passage, a conical tower of stone, faced with blue bricks, was erected in **1876. Carries a white** light, revolving every 90 seconds.

Castle **Island, in the same** channel, is marked by a fixed white light; conical tower, with three bands of brick, 114 feet; erected 1868.

Turk Island.—The tower is of iron, 60 feet; erected **in 1845.** White light, revolving every minute.

Inagua Great Island.—Conical white **tower, 114** feet high, **with white** light, revolving every minute, visible 17 miles.

Upon *Cape Pembroke*, in the lonely and sterile Falk-**land** Islands, a light was exhibited on December 1st, 1855, for the benefit **of** vessels bound round Cape Horn and entering Fort William. The tower, 60 feet **high, is** of iron, and was constructed in England by **Messrs. Wilkins for about £2,400. Has a fixed white** light, catoptric, visible **for 12 miles.**

Cape Race, Newfoundland, is the south-east corner of that island of fogs and fishers, and notable as being generally the last land left and the first gained by vessels engaged in carrying on the trade between Great Britain and the United States. Its summit is crowned by an iron lighthouse, designed by Mr. Alexander Gordon, constructed in England, and sent out to be put together under the superintendence of a resident engineer. The circular tower rises from the centre of the keeper's dwelling to a height of 50 feet, and is painted in red and white vertical stripes. Cost, £7,358. Seal oil is used to feed the light—a fixed white light, visible for 15 miles.

Another of Mr. Alexander Gordon's iron lighthouses, 40 feet high, was erected in 1857, and lighted on the 1st January 1858, on *Breaksea Island,* at the entrance of King George Sound, Western Australia. It has a fixed white light, third order, dioptric, at an elevation above high water of 383 feet.

Point King, the northern bluff of the narrow entrance to Princess Royal Harbour, King George Sound, is the site of a small wooden lighthouse (with keeper's dwelling attached, looking very much like an English cottage), 17 feet in height, which shows a fixed white light, dioptric, fifth order, over 7 miles. Cost, together with the lighthouse on Breaksea Island, £3,796.

Cape of Good Hope.—Table Bay, on the south-west

side of the bold peninsula to which the early navigators gave so auspicious a name, is well lighted. There are lighthouses, for example, on Robben Island,* Green Point, and Monilli Point, besides lights on the breakwater and Prince Alfred Pier. To the south, on *Cape Point,* in lat. 34° 21', stands an iron lighthouse, 30 feet high, designed by Mr. Alexander Gordon, at a cost of about £4,500. It has a revolving white light, visible for 12 seconds every minute, and commanding the unusually extensive range of 36 miles. The focal plane of this lighthouse is 816 feet above the sea, so that it is one of the loftiest lighthouses in the world. It was first lighted on the 1st May 1860.

On the southernmost of the *Roman Rocks,* in Simon's Bay (eastward of the Point), an iron tower, 48 feet high, was erected in 1861, at a cost of about £7,600. The lantern and illuminating apparatus cost £1,025 additional. A revolving white light, visible for 12 seconds every minute, is shown here.

In 1848 a lofty lighthouse was erected on *Cape Agulhas.* It measures 100 feet from base to vane ; is a circular tower painted with alternate bands of red and white ; and exhibits a fixed white light, with a range of 18 miles in fair weather.

---

* "Robben Island looked like a dun-coloured hillock as we shot past it within a short distance, and a more forlorn and discouraging islet I don't think I have ever beheld. When I expressed something of this impression to a cheery fellow-voyager, he could only urge in its defence that there were a great many rabbits on it. If he had thrown the lighthouse into the bargain, I think he would have summed up all its attractive features."—LADY BARKER.

Lady Barker, in her lively record of South African experience, remarks that this part of the coast is well lighted (it is very dangerous), and adds :—" It was always a matter of felicitation at night, when every 80 miles or so the guiding ray of a lighthouse shone out in the soft gloom of a starlight night. One of these lonely towers stands more than 800 feet above the sea level, and warns ships off the terrible Agulhas Bank."

Cape St. Francis (1½ mile west of), lat. 34° 12'; flashing light, 20 seconds; white, with red sector; dioptric, second order; visible 16 miles; cylindrical white tower, 91 feet high; erected in 1878.

Cape Recife, lat. 34° 2', long. 25° 42'; revolving light every minute; white, with red sector; visible 15 miles; tower 80 feet high, painted with four horizontal bands of red and white alternately ; erected in 1850.

Bird Island, Algoa Bay, lat. 33° 50', long. 26° 17'; fixed red light, visible 14 miles; square tower, 72 feet high; erected 1852.

Natal, on bluff south side of harbour, lat. 29° 53', long. 31° 4'; revolving white light, visible 24 miles; conical iron tower, painted white, 81 feet high, with focal plane of light 282 feet ; erected in 1867.

Isle Fouquets, south entrance of Grand Port, island of Mauritius, lat. 20° 23' N., long. 57° 47' E.; fixed white light, visible 16 miles; gray tower, with lantern painted red, 84 feet in height; erected in 1864.

Zanzibar, *Mungopani*, on the cliff northward of the

village, lat. 5° 37', long. 39° 11'; fixed white light, visible **12 miles**; square white tower, 90 feet in height; erected in 1886.

Aden, Murshigh Cape, lat. 12° **45'**, long. 45° 3'; fixed white light, visible **20** miles, first order, dioptric; dark gray tower, 85 feet high; focal **plane of** light 244 feet above the sea; erected in 1867.

Red Sea, Dædalus Shoal, **lat. 24° 55'**; fixed white light, visible **14** miles; on **open** iron-work, **70** feet high; erected **1863**. **There is a** similar structure on the *Ashran Reef*, Jubal Strait, lat. 27° 47'; **140 feet high, with a** revolving white **light**; erected 1862.

Red Sea, Point Zafarana, **lat. 29° 6'**; fixed white light; circular stone tower, 82 feet high; erected 1862.

Tracing the west coast of **Hindustan, we** notice among the more important lighthouses :—

Karachi, **lat. 24° 47'**; revolving white light; circular white tower, 52 feet high; erected **1848**.

Bet Harbour, in Gulf of **Kutch, lat. 22° 29'**; fixed white light, with red **sector**; circular stone tower, **65 feet**; lighted 1876.

Perim Isle, Cambay Gulf, lat. **21° 36'**; fixed white light, visible **15 miles**; circular tower of brick, 78 feet high; erected 1851.

Tapti, mouth of **Surat River, lat. 21° 5'**; fixed white light, **visible 15 miles**; circular tower, painted with **three belts of red and white**, 90 feet high; **erected in 1852.**

Bombay, on sunk rock in lat. 18° 49′; occulting white and red light, visible 14 miles; tower 95 feet high; erected 1884.

Bombay, South-West Prong, lat. 18° 53′; white light, flashing, visible 18 miles; tower 146 feet high, painted with alternate bands of white, red, white, and black; erected in 1874.

Bombay, Kennery Island, lat. 18° 42′; fixed white light, with red sector, visible 19 miles; octagonal tower rises from the centre of a flat-roofed house, 75 feet high; erected in 1867.

Alipec, lat. 9° 30′; white light, revolving every minute, visible 17 miles; white tower, 115 feet high; erected in 1862.

Point de Galle, Ceylon, lat. 6° 1′; fixed white light; circular tower of iron, painted white, designed by Mr. Alexander Gordon, 80 feet high, erected in 1848.

Great Basses, on the north-west rock, lat. 6° 10′; red light, revolving every 45 seconds; granite tower, 127 feet high, with two galleries, one at a point 30 feet above base, another under lantern; designed by Alexander Gordon and Sir James Douglass; completed (after a long delay) in 1873.

Little Basses Rocks, lat. 6° 23′; white light, with two flashes every minute at intervals of 15 and 45 seconds, visible 16 miles; granite tower, 127 feet high, designed as above; erected in 1878.

Tuticorin, on the coast of Coromandel, lat. 8° 47′;

fixed white light, visible 14 miles; tower of 91 feet, painted brown, lantern white; erected in 1845.

Coringa Bay, on southern part of Hope Island, lat. 16° 49′; fixed white light, visible 14 miles; tower 94 feet high; erected 1827.

False Point, entrance to Mahanuddy River, lat. 20° 20′; white light, occulting, eclipsed for four seconds every half minute; tower 132 feet high, painted reddish brown, with large white star in the centre; erected 1838.

Saugor Island, Middleton Point, lat. 21° 39′; fixed white light, with flashes every 20 seconds, visible 15 miles; lighthouse 76 feet high, painted with bands of red and white; erected in 1821.

Kutabdia (west part of island of), Bay of Bengal, lat. 21° 52′; fixed white light, visible 12 miles, lighthouse 111 feet high, lower part painted gray; erected, 1846.

ALGUADA REEF LIGHTHOUSE.

Alguada Reef, lat. 15° 42′, long. 94° 12′; white light, revolving every minute, and visible 20 miles; noble tower of granite, 160 feet high, one of the highest in the world; erected in 1865.

China Bakir, at end of flat, in lat. 16° 17′ N., and long. 96° 11′ E.; fixed white light, with flashes, visible 15 miles; light-room and lantern upon screw piles (Mitchell's system); erected in 1869.

Rangoon River, on the east side of the entrance, lat. 16° 30′; fixed white light, visible 12 miles; screw-pile lighthouse, 102 feet high; erected in 1876.

Double Island, north point, lat. 15° 52′; fixed white light, 19 miles; tower of masonry, 75 feet high; erected 1865.

Table Island, Great Coco Group, Andaman Islands, lat. 14° 12′; fixed white light, visible 22 miles; circular lighthouse, painted with alternate belts of white and red, 91 feet high; erected in 1867.

Pulo Brasse (Achi Head), lat. 5° 45′; white light, revolving every minute, visible 30 miles; tower 120 feet high, painted white up to 98 feet, and then red; focal plane elevated 525 feet above the sea; erected in 1875.

Muttra Head, north-west point of Penang, lat. 5° 28′, long. 100° 10′; white light, with flashes; tower of gray granite, 45 feet from base to vane; focal plane of light 795 feet above the sea; erected in 1883.

Malacca, the old monastery on St. Paul Hill, famous in connection with the labours of Francis Xavier, lat. 2° 12′; fixed white light, visible 15 miles; square tower, 90 feet high; first lighted in 1849.

Horsburgh, or *Pedro Banca*, on summit of low rock lat. 1° 20′ S.; revolving light every minute; circular white tower, 106 feet high; erected in 1851.

Sumatra, Pulo Bojo, south-west end of island, lat. 0° 38′ S.; white light, with two flashes every 30 seconds; the lighthouse, a strange, sixteen-sided building, white, 197 feet high, and centre of lantern 361 feet above the sea; erected in 1885.

Java, on *first point,* Sunda Strait, lat. 6° 44′ S.; white light, with flash of six seconds, followed by 24 seconds of darkness, visible 23 miles; lighthouse 131 feet from base to vane; first lighted in 1877.

Java, on *fourth point,* lat. 6° 4′ S.; fixed white light, visible 20 miles; lighthouse 177 feet in height; erected in 1855.

Java, on *Flat Cape,* lat. 5° 59′ S.; white light, shows in rapid succession three flashes of two seconds each, separated by three seconds of darkness, and followed by an eclipse of 18 seconds, visible 23 miles; lighthouse sixteen-sided; focal plane 213 feet above high water; erected in 1880.

Sourabaya Strait, Sembilemgan, lat. 7° 4′ S.; fixed white light, visible 19 miles; lighthouse painted white, 164 feet from base to vane; erected in 1882.

Gaspar Strait, Shoalwater Island, lat. 6° 19′ S.; fixed white light, visible 20 miles; white, sixteen-sided building, 215 feet from base to vane; erected in 1883.

Gaspar Strait, Langwas Island, lat. 2° 32′; intermittent white light, fixed for 60 seconds, eclipsed 25, flash 10 seconds, eclipsed 25, visible for 10 miles; sixteen-sided building, 215 feet high; first lighted in 1883.

CHINA.

Breaker Point, beyond **Hong Kong,** lat. **22° 56';** white and red light, occulting, 10 seconds; circular tower, with black and white bands, 120 feet high; erected in 1880.

Formosa Island, South Cape, lat. **21° 55';** fixed light, white and red, visible 20 miles; circular tower, 71 feet high; erected in 1882.

Dodd Island, lat. **24° 26' N.;** white and red occulting light, seen for 26 seconds, eclipsed for four; circular tower, white, and 79 feet high; erected in 1882.

Ocksen Island, lat. **24° 59';** revolving light, visible 24 miles; lighthouse tower black, 64 feet high, keeper's dwelling and enclosure wall white; erected in 1874.

Middle Dog Island, near Min River, in lat. **25° 58' N.;** fixed white light, with flashes every half minute, visible 23 miles; lighthouse tower white, circular, 64 feet high; erected in 1872.

Sha-lin-tien Island (Tsao-fri-tien), lat. **38° 56';** fixed white light; octagonal tower, brick and stone, 45 feet high; erected in 1886.

JAPAN.

Tebosi Sima Island, lat. **33° 41';** fixed white light, visible 20 miles; octagonal tower, 57 feet high; erected in 1875.

Tsuno-Sima, west point of, lat. 34° 21′ N.; white light, flashing every 10 seconds, visible 18 miles; circular tower of granite, 100 feet high; erected in 1876.

Siwo Msaki, south point of province of Kii, lat. 33° 26′; fixed white light, visible 20 miles; circular white tower, 75 feet high; erected in 1873.

Omai-Saki, south part of cape, lat. 34° 36′; white light, revolving every half minute, visible 19 miles; circular white tower, 72 feet high; erected in 1874.

Mikomoto (*Rock Island*), lat. 34° 34′; fixed white light, with red sector, visible 19 miles; tower painted white, with two black bands, 72 feet high; erected in 1871.

Inu-Bo-Ye-Saki, south-east extremity of promontory, in lat. 35° 43′; white light, revolving every half minute, visible 19 miles; circular white tower, 103 feet high; erected in 1874.

Siriya Saki, on the cape, lat. 41° 26′; fixed light, white, visible 18 miles; circular building, painted white, 94 feet high; erected in 1876.

Gulf of Tartary, Retchnoi Island, entrance to Suifun River, lat. 43° 16′; fixed white light; quadrangular building, 106 feet high; erected in 1885.

AUSTRALIA.

Port Walcott, Reader Head, lat. 20° 39′ S., long. 117° 13′ E.; fixed white light; erected in 1881.

Rottnest Island, lat. 32° ; revolving every minute, **visible 22** miles ; tower of stone, 64 feet ; **erected** in **1850.**

Swan **River,** *Arthur Head,* **lat. 32° 3′ S.** ; long. **115°** 45′ E. ; fixed white light, visible 16 miles ; circular lighthouse, 71 feet high ; erected **in 1851.**

King George *Sound, Breaksea Island ;* fixed white light, visible **24** miles ; iron tower, designed by Gordon, 43 feet high ; erected in 1858.

Spencer Gulf, **on** *Tipara Reef,* lat. 34° 3′ ; white light, revolving every half minute, and visible 20 miles ; erected on iron piles in 1877.

Kangaroo Island, Cape Flinders, lat. 35° **46′** ; red **and** white light **alternately,** revolving **every** half minute ; white visible 30, and red 15 miles ; square tower of iron, designed **by** Alexander Gordon, **60** feet high ; erected in 1858.

Troubridge **Shoal, St. Vincent** *Gulf,* **lat.** 35° 7′ ; white light, revolving **so as** to show bright for 24 **and be** eclipsed **for 36** seconds ; **iron** light-tower, **70** feet, **painted in alternate** stripes of white and **red, 20** feet **wide ; designed by** Alexander **Gordon, and erected** in 1856.

Kangaroo Island, **Cape** *Willoughby,* lat. 35° 51′ ; **white light,** revolving every **90** seconds, visible for **24 miles ; white tower,** 75 feet **high, erected in 1852.** " We sighted **Kangaroo** Island," **say the authors of** " What we saw in Australia," " about one o'clock P.M., and **by** three were running almost close **under its** long level line of cliffs. A lighthouse, and one or two minute farmhouses, were the only signs of man's presence that we could discern, and not a tree was to be seen."

Cape Northumberland, lat. 38° 3'; revolving white and red light with flashes; visible 20 miles; iron lighthouse, designed by A. Gordon, painted with three broad bands, white, red, white; erected in 1859. "Towards evening we saw at some little distance inland Mount Gambier, which rises in a curious volcanic district on the eastern borders of South Australia; and somewhat later we passed Cape Northumberland, with its brilliant red and white revolving light, at the western point of Discovery Bay."

King Island, Bass Strait, lat. 39° 36'; fixed white light, visible 24 miles; circular white tower, 145 feet from base to vane; erected in 1861.

Shortlands Bluff, two miles within the entrance of Port Philip, lat. 38° 16'; fixed lights, upper white, lower white and red, visible 17 miles, 14 and 10 miles; blue stone building, 81 feet high; erected in 1863.

Cape Schanck, on summit to the south, lat. 38° 10'; fixed white light, with flashes (two minutes); circular stone tower, painted white, about 70 feet high; focal plane of lantern 328 feet above the sea; erected in 1859.

Wilson Promontory, south-east point, lat. 39° 8'; fixed white light, visible 24 miles; circular stone building, **70** feet high, with the focal plane of lantern 342 feet above the sea; erected in 1859.

Gubo Island, south-west of Cape Howe, in Bass Strait, lat. **37°** 35'; fixed white light, visible **20** miles; fine circular tower of gray granite, 156 feet in height; erected in 1856.

Green Cape, lat. **36°** 17'; white, flashing, one minute, visible **19** miles; lighthouse 68 feet high; erected 1883.

Jervis Bay, lat. 35° 9'; light alternately white, red, and green; tower painted white, 61 feet high; erected 1860.

Port Jackson (for Sydney), outer *South Head*, lat. **33°** 51'; white light, revolving every minute, visible **21** miles; circular stone tower, **76** feet high, with its focal plane 344 feet above the sea; erected in 1817; was lighted by electricity in 1883.

Port Jackson, inner *South Head*, or *Hornby Light*, on edge of cliff, which is about **50** feet in elevation; white light, fixed, visible **14** miles; tower painted with red and white vertical stripes, **50** feet high; erected in 1858. "About a mile before we reached the entrance to Sydney Harbour we passed a slight inward curve in the rocky cliffs, the scene of a terrible shipwreck which took place in August 1857. The captain of the *Dunbar*, a vessel containing amongst its passengers many leading colonists returning home from Europe, mistook in the dark this curve for the expected channel, and steered his ship full upon the rocks. It struck violently, and at once became a total wreck.

Only one man on board escaped with life. He was cast upon a ledge of rock high above the sea, and remained there till the next day, when he was discovered and rescued from his perilous position.

"An opening between the precipitous cliffs, called the North and South Heads, gives access to Port Jackson, so named after one of Cook's sailors, who discovered the entrance. Cook himself is said never to have sailed into the harbour. The area of Port Jackson proper, now generally called Sydney Harbour, measures nine square miles, and that of Middle Harbour, one of its arms, three square miles, while the coast-line of the whole is fifty-four miles in length. At the time of the wreck of the *Dunbar* there was but one lighthouse at the entrance, and that was upon the South Head. Under the supposition that the captain may have been misled by the single light, another has since been placed there, in the hope of preventing the repetition of so terrible a catastrophe."

Port Stephens (south side of entrance), lat. 32° 45'; white and red light revolving alternately every minute, visible 17 miles; circular stone tower, 73 feet high; erected 1842.

Sandy Cape (on the summit of Great Sandy Island), lat. 24° 43' S.; revolving white light every two minutes, visible 27 miles; iron tower 99 feet high, painted white; the focal plane of light elevated 400 feet above high-water mark.

TASMANIA.

Banks Strait, Goose Island, lat. 40° 19'; **fixed white light**, visible 14 miles; tower 74 feet high, **upper part red** and lower part white; erected 1846.

Banks Strait, **East Point**, **lat. 40° 44'**; **white** light, revolving **every** minute, visible **15** miles; circular **tower, 74 feet high;** erected **1845.**

NEW ZEALAND.

Poteaux Strait, Dog Island, **lat. 46° 40';** white light, revolving every half minute, visible **18** miles; tower **118** feet high, painted **with** black, white, **and black** bands; **erected in 1865.** Illuminating apparatus catadioptric, **first order. This is the only** first-class lighthouse of **which New** Zealand as yet can boast.

CHAPTER X.

FLOATING LIGHTS, OR LIGHTSHIPS.

WE have seen that the inner line of coast defences which man has devised for the protection of those that go down to the sea in ships is formed by our lighthouses—lighthouses on shore and rock, on the projecting headland and the outlying reef. But it is obvious that these are inadequate to meet every requirement of the mariner; that they can be of no service, for instance, in the navigation of the shoals and sandbanks which frequently lie at a distance of twenty or thirty miles out at sea, and being submerged at high water, present a very formidable, because not easily detected, danger. We may point, for an example, to the Goodwin Sands, off the shores of Kent; a fatal spot, which has been the destruction of many a goodly vessel and her crew, and for centuries has borne the burden of memories of sad calamities. Long was it considered impossible to supply any certain and permanent warning of its hidden perils. No tower of masonry or iron could be erected on a foundation so unstable and treacherous; and it seemed as if this one wild waste must for ever remain exposed to the pitilessness of stormy seas which strewed it with frequent wrecks. At last, however, the idea occurred to an inventive mind of stationing floating lights at this and similar places; that is, of substituting lightships for lighthouses in positions where permanent structures could not be employed.

Robert Hamblin was a respectable barber of King's Lynn, who had married the daughter of a shipowner, and in due time had become owner and master of a vessel, with which he embarked in the Newcastle coal trade. Accident introduced him to a man of remarkable ingenuity named David Avery, whose career had been marred by his great poverty. Discovering that he had conceived the idea of floating lights, he advanced the money necessary to carry it out, and in 1732 the two men established at the Nore a lightship, and proceeded to levy tolls on passing ships for her maintenance. Their illuminating apparatus, however, was of the simplest construction—a small lantern provided with flat-wick oil lamps, but unaided by optical apparatus of any kind.

Though compelled to own that the new lightship was of great assistance in the navigation of the Thames, the Trinity House did not fail to regard the action of Hamblin and Avery as an encroachment on their privileges; and when they announced their intention of stationing a similar vessel among the waters of the Scilly Islands, they laid a complaint before the Board of Admiralty. As it was unable or unwilling to interfere, the corporation next appealed to the Crown, representing that it was illegal for any private individual to levy a tax on the mercantile marine; and they moved with so much energy as to obtain the issue of a royal proclamation prohibiting the exaction of dues in respect of the lightship at the Nore. Avery, on experiencing this severe check to his plans, appeared in person before the Board, and proposed terms of agreement. He stated that he had expended on the Nore Lightship a sum of £2,000; and he offered that all right and title to it should devolve upon the Trinity Corporation, provided that he was allowed to levy the tolls by his representatives and heirs for a period of sixty-one years, on payment of £100 yearly.

The Nore lightship had proved so beneficial that, at the urgent request of all engaged in the coasting trade on the east, a lightship was moored in 1736 at the Dudgeon Shoal, at the entrance of the Wash. For a long period nothing more was done, but in 1788 it was resolved to station a floating light on the Owers Shoal, off the Surrey coast. In 1790 one was placed close to the Newarp Sand, off the Norfolk coast, and in 1795 a vessel was anchored to the north-east of the Goodwins. Up to this date the lightship had shown only *two* lanterns, set horizontally on a cross-yard on the mast; but from the Newarp ship were first suspended three lanterns in a triangle. The Sunk Lightship, near to the entrance of the Thames (which is now so copiously lighted), dates from 1798; the Galloper from 1803; and the Gull Stream, on the Goodwins, from 1809. The shore waters of England and Wales are now lighted by fifty-seven of those useful vessels. On the Scottish coast, however, only four are employed, and on the Irish only eleven, because these coasts are lined with iron-bound cliffs, which do not form shoals or sandbanks in the adjacent waters, and lighthouses, therefore, on prominent points afford all the guidance and protection needed.

The rude kind of light which we have described was in use on board our lightships until, in 1807, the late Mr. Robert Stevenson, the eminent Scottish engineer, designed a larger lantern to surround the mast of the vessel, capable of being lowered when the light required trimming, and of being raised when the light required to be shown. When the catoptric illuminating system was introduced, an improvement was effected in the light by the employment of Argand burners, assisted by paraboloidal silvered reflectors, burners and reflectors alike being properly "gimballed" to insure the horizontal direction of the beam of light during the motion of the vessel.

In 1872 the Trinity House increased the dimensions of their lanterns and reflectors for floating lights, the lanterns from six feet to eight feet in their diameter, with cylindrical instead of polygonal glazing, and the reflectors from twelve inches to twenty-one inches aperture, thus securing a tenfold increase in the intensity of these lights. The cylindrical lanterns now in use are large enough to admit the entrance of the lamplighter to trim the lamps. Further improvements have been made, with the result that some of these lights have an intensity in the beam of about twenty thousand candles. In 1875 the first group-flashing floating light, showing three successive flashes at one-minute periods,* was installed on board a new lightship moored at the Royal Sovereign Shoals, off Hastings, and this class of floating lights has since been considerably extended. In a few cases the dioptric system has been adopted for light-vessels; but, considering the special conditions under which they act, the catoptric has on the whole been found more efficient and satisfactory.

The English lightships are invariably painted *red*, and the Irish *black*, with the name in large white letters on both sides. By day they carry at the masthead a large wooden sign, either circular, semi-circular, triangular, or otherwise in shape. The mast is encircled by the lantern, which is lowered during the day. Sometimes the ship has two and even three masts, and, of course, a lantern for each.

The lantern contains a certain number of lamps and reflectors, each hung upon a gimbal, so that it moves

* In 1887 altered to three quick flashes every forty-five seconds,

freely in all directions, and by virtue of its own gravity maintains a vertical position, however much the
vessel may pitch or roll. For a fixed light, they are
adjusted so that a beam of light may uninterruptedly
be diffused all round; for a revolving light, on each
face (three or more) of the iron framework (which
rotates on a spindle) are set three, or five, or seven

A LIGHTSHIP.

lamps and reflectors, as the case may be. By ingenious clockwork mechanism the revolution of the
framework is regularly kept up, and each face brings
round its lights at a fixed interval. The number of
lamps used on board a lightship varies from nine to
twenty-four, each of which consumes in a year about
thirty-six gallons of rape or of colza oil (mineral oils
are prohibited as too dangerous).

That it is of primary importance a lightship should
be moored securely, the reader will at once perceive.
If she broke adrift, not only would the safety of the
crew be endangered, but her warning light would be
absent from its station possibly when most wanted.
Some twenty-five years ago such a mishap was not
of infrequent occurrence, but this was due to defects
which experience has revealed and increased knowledge removed. The links of the mooring cable now
in use are made of iron one-and-a-half inch diameter.
The chains are specially manufactured, and before they
are accepted the iron of each link is tested to bear a
tensile strain of twenty-three tons per square inch of
the original area, while the whole cable, when tested
for the welding, is required to bear a pressure of
sixteen tons per square inch of each side of the link.

The chains are made in fifteen-fathom lengths, with
a swivel to prevent kinking in the centre of every
alternate length, and are joined by shackles, all of
which are required to bear the same tests as the cable
links. To each ship are supplied from two hundred
and ten to three hundred and fifteen fathoms. The
method of mooring is either by a single mushroom or
Martin anchor (weighing two tons), in which case the
ship swings round as the tide changes; or two mushroom anchors are connected by a length of two-inch
ground chain, in the centre of which are a ring and
swivel attached to the one-and-a-half inch cable
chain, and in this case the vessel swings in a very

limited area, suitable for a narrow fairway or channel. Each lightship **also** carries two bower anchors for use in emergencies, and an additional length of one hundred and fifty fathoms of one-and-a-half inch chain.

"**But** although it is of **the utmost** importance," says Mr. Edwards, "that every precaution be taken to **obtain the** very best quality **of** materials, yet it **is chiefly** by the skilful management of the mooring cable that a lightship is enabled to ride **out** the fiercest storms **in** safety. **With a** smooth sea **a short** cable is sufficient, but when **the** waves run high **it is** necessary to pay **out a long** scope of chain, so that **the ship** may ride **over** the highest crests and plunge down **into the lowest** depths **of the** trough **of the** sea. It is also necessary **to** have **a** very great deal more cable out than is actually required to enable the vessel to surmount the highest waves; **she must** never be allowed to go **to** the end **of her** tether **and** pull directly upon her mushroom: as the vessel **rises she** takes as much chain **as she** requires, but **still must have a** considerable quantity **on the** sea-bed. **This** surplus cable by its own weight acts **as** a spring, and entirely prevents any direct jerking or straining at **the** mushroom. The experience of years **has** educated the officers of our lightships to regulate the scope of cable paid out to the necessities of the occasion. The constant rise **and** fall of the cable and the swinging round of the vessel with **the** tide are at times the cause of strange entanglements, and **it is by no** means an uncommon duty for one of the steamers of the service to go out to ' clear the moorings ' of a lightship."

When first seen, and especially if seen from a distance, a lightship closely resembles, during the day, an ordinary vessel—a vessel of one hundred and sixty to one hundred and eighty tons, measuring eighty to ninety feet from stem to stern. She is generally built of wood, or of composite construction, fastened and sheathed with a patent metal. A few vessels have been built of iron, but these are found at the end of one or two years to require removal from the station, docking, and the external submerged portions of the hull to be cleaned and painted. On examining our lightship from a nearer point of view, we discover some marked peculiarities. Though she floats, she does not move ; other ships represent motion, she represents immobility. Her form varies according to locality,—for instance, in Ireland the hull is more elongated than in England,—but always the primary consideration is, how best to secure resistance to the force of the winds and waves.

We believe there is no instance on record of the crew of a lightship having, under any circumstances, voluntarily abandoned their position. With a steadfast courage and coolness worthy of the highest praise, they brave the fury of the most violent tempests, though unsupported by that excitement of constant action which the sailor experiences out in the open. Should it so happen that their vessel is driven from its moorings by the stress of the gale, they hoist a **red signal and fire a** gun, **and** generally assistance **is** immediately forthcoming.   **As** it is necessary to be prepared for every contingency, a spare vessel is always kept in readiness at the central station of each district.   Owing to **the** telegraphic network with which **our** shores are now surrounded, an accident is speedily made known to the authorities, and often before **sunset the** reserve ship, **towed by** a powerful steam-tug, **replaces** the one which has temporarily been disabled.

The crew of a lightship consists of eleven men,—a master, **a** mate, three lamplighters, and six seamen. Of these, seven only **are** on board.   The master and **mate** have alternate months afloat and ashore; **the** others have each two months afloat and one month ashore.   **When** ashore the men are employed upon certain duties at **the** district **depot.**   Once a month every lightship is visited by the official steamer, which carries out the men whose month ashore has expired, and **brings** back those whose turn **it is to** be relieved from sea-duty.   On this occasion, moreover, fresh supplies of fuel, oil, water, and provisions are put on board. Experience has proved that continuous service afloat is too much for the moral and physical forces of human nature to endure.   **The** crushing pressure of monotony, the unchanging spectacle of rolling waters, the ceaseless murmur of the winds, ocean's everlasting voices— these necessarily exercise a depressing influence on both body and mind.   Even with the relief of the monthly sojourn ashore, the life is so uniform, and, on the whole, so devoid of active interest, that it is wonderful men can be found to submit to it, and the crews of our lightships may fairly be ranked among the curiosities of civilization. We know not whether the confession of an old lightship hand would be adopted by all his comrades—that when he was ashore he dreamed constantly of the sea, and when he was afloat, could dream of nothing but the land; but this at least is certain, that they submit to discipline with the most exemplary obedience, and always exhibit the utmost patience and cheerfulness. In fact, whoever visits one of our floating lights must needs be impressed by the fine appearance of its crew. Weather-beaten and sun-tanned, they are models (says Esquiros) of English seamen —frank, self-reliant, unassuming, obedient, vigorous, agile, and resolute. Their wages average fifty-five shillings per month, with rations; the master receives £80 per annum.

The lightships, besides lights, are supplied with guns and gongs, rockets, and, in some cases, with fog-horns. The following regulations apply to those moored off the coasts of England and Ireland:—

A white light is exhibited from the forestay, at a height of six feet above the rail, to show in which direction the vessel is riding, when at her station.

When she is driven from her proper position to one where she is of no use as a guide to shipping, she will not exhibit her usual lights, but a fixed red light both at stem and stern, and a red flare every quarter of an hour. By day, the balls or other distinguishing mast-head marks will be struck.

If from any cause she is unable to show her usual lights whilst at her station, the riding light only will be displayed.

The mouths of fog-horns in light-vessels are pointed to windward; also those on the land to seaward.

When, from any lightship, a vessel is seen standing into danger, a gun will be fired and repeated until observed by the vessel; also, the two signal flags "S. D." of the Commercial Code, "You are standing into danger," will be hoisted and kept flying until answered.

In England and Ireland, whenever a light-vessel or other craft is anchored to mark the position of a wreck, the top-sides will be coloured green, and she will be further distinguished, by day, by three balls placed on a yard thirty feet above the sea, *two balls (vertically) on the side on which navigating vessels may safely pass,* and one on the other; and by night, by three fixed *white* lights similarly arranged, and with the same meaning. Those marking vessels, when so employed and fitted, will not show the ordinary riding light.

Note.—The firing special rockets (of little sound, but of great brilliancy) immediately after a gun from a light-vessel will denote the need of assistance from the shore.

The chief danger to which a lightship is exposed

arises from collisions, and unfortunately these are of frequent occurrence. "Some years ago the Tongue Lightship, at the entrance of the Thames, was run into by one of the steamers which trade regularly into and out of the river, and was cut down to the water's edge, so that she sank almost immediately, the master and crew being saved with difficulty."

Chinese gongs, about two feet in diameter, sounded at short intervals, have long been the recognized standard fog-signal of lightships, owing perhaps to their peculiar distinctive sound. At short distances this is very serviceable, but, like the sound of a bell, it is soon dissipated, and has no considerable range. Many of our ships are now provided with powerful sirens or whistles, sounded by compressed air or steam. Owing to their positions (generally some miles from the shore) they are found, owing to the uncertain range of fog-signals, to be more efficient aids to navigation than those installed at shore stations.

On board all lightships the life of the crew is much the same. At dawn, on Sunday, the lantern is lowered, and the lamplighter cleans and prepares his lamps for the next night's work. At eight o'clock every man must be on the alert; the hammocks are hung up, and breakfast is served. Afterwards, the men wash, and proceed to don their uniform, of which they are very proud, for the arms of the Trinity House are blazoned on the buttons. At half-past ten they assemble in the

cabin, and the captain or mate reads prayers. The **day**
is spent quietly, with as much **cessation from labour as**
the conditions of lightship **service permit**; and at sunset the lighted lantern is **hoisted, and the crew again**
meet together for prayer and the reading of the
Scriptures. Apart from the morning and evening
services, **most of the** week **days bear a close resem-**
blance to the Sunday. The two **cleaning days are**
Wednesday and Friday, when **mop and bucket are in**
active requisition. To watch over and maintain in
perfect order the illuminating apparatus; to keep
watch on deck; to record seven times in every twenty-
four hours the direction of the wind and the state of
the weather; to test the continued efficiency of the
mooring-chains—such is the almost invariable circle of
their daily occupations. Their leisure, which is by no
means inconsiderable, they employ in reading. A
library is always kept on board, and the books are cir-
culated from hand to hand and ship to ship.

Mr. Ballantyne, who once spent a week on board the
Goodwin Sands Lightship, has published a lively narrative of his experiences. "That curious, almost ridicu-
lous-looking craft," he writes, "is among the aristocracy
of shipping. Its important office stamps it with no-
bility. It lies there, conspicuous in its royal colour,
from day to day and year to year, to mark the fairway
between old England and the outlying shoals, distinguished in daylight by a large ball at its masthead,
and at night by a magnificent lantern, with Argand

lamps and concave reflectors, which shoots rays like lightning far and wide over the watery waste; while in thick weather, when neither ball nor light can be discerned, a sonorous gong gives its deep-toned warning to the approaching mariner, and lets him know his position amidst the surrounding dangers."

He goes on to say of the wave-tossed craft that it is "an interesting kingdom in detail." The visitor, standing abaft the one mast, sees before him the binnacle and compass and the cabin skylight. On his right and left the territory of the quarter-deck is seriously circumscribed and the promenade much interfered with by the ship's boats, which, like their parent, are painted red, and do not hang at the davits, but, like young lobsters of the kangaroo type, find shelter within their mother when not at sea on their own account. Near to them stand two signal carronades. Beyond the skylight rises the bright brass funnel of the cabin chimney, and the winch by means of which the lantern is hoisted. Then come another skylight and the companion hatch about the centre of the deck. Just beyond is the most important part of the vessel—the lantern-house, a circular wooden structure, about six feet in diameter, with a door and small windows, which encloses the lantern, the beautiful piece of mechanism for which the lightship, its crew and appurtenances, are maintained. Right through the centre of this house rises the thick, unyielding mast of the vessel; and the lantern, which is only just a little smaller than its house, surrounds the mast and travels upon it. It is, of course, connected with the rod and pinion by means of which, with the ingenious clock-work beneath, the light is made to revolve and flash once every third of a minute.

Such is a general survey of the Gulf Stream light-vessel, and other light-vessels resemble it in all leading particulars. *Ex uno disce omnes.*

CHAPTER XI.

LANDMARKS, BEACONS, BUOYS, AND FOG-SIGNALS.

TO complete our account of the aids to navigation devised by man's thought and skill, we must bring under review some works of less ingenuity and pretension than our lighthouses and lightships, and of less importance, but still of unquestionable utility. We refer to those which are known as landmarks, buoys, and beacons.

As to landmarks and beacons, we may say that, in general terms, they include every kind of terrestrial object which assists the mariner in steering his course. For this purpose, the spire of a church, the tower of a ruined castle, a windmill on its breezy height, an isolated tree, or even a rock if of characteristic configuration, may be useful, and landmarks of this description will be found laid down with much care in our seamen's charts. Of those which are artificial and specially erected as guides to navigation, we may cite a famous example. We pass over the so-called Pillars of Hercules, which legend placed on either side of the

Strait of Gibraltar, at Calpe and at Abyla, because
their existence may reasonably be doubted; but a word
or two must be given to the tall and shapely shaft of
granite which overlooks Lake Mareotis and the modern
city of Alexandria. Its total height, including base
and capital, is 98 feet 9 inches. For generations it has
been known as " Pompey's Pillar," but an inscription on
its pedestal informs us that it was erected by Pompius
or Publius, a Roman prefect of Egypt, in honour of
the Emperor Diocletian, " The Invincible," and to com-
memorate the deliverance of Alexandria from the in-
surgent forces of the pretender Achilleus in A.D. 297.
It now serves as a notable landmark for ships leaving
or entering the port of Alexandria.

As commerce developed its proportions, and discovered new channels for its development, it became
necessary to multiply along every coast the beacons
which performed by day the same useful part as that
played at night by the signal-fires. According to
Coulier, we owe to the Etruscans the invention of that
system of landmarks which, after long neglect, has,
within the last half-century, been revived and carried
out on an extensive scale. Where no natural landmarks are available, or no existing buildings, we raise
at suitable points small structures of timber or of
masonry, painting them of a brown colour, if they
stand defined against the sky, as on the summit of a
lofty hill, or white, if they stand projected upon the
sea. But where permanent beacons are impracticable,

COLUMN AT ALEXANDRIA KNOWN AS POMPEY'S PILLAR.

as in estuaries, **rivers, and** narrow channels, we substitute *buoys*—that is, floating frameworks of wood **or** of **iron, with or without a** ball, and painted of different **colours.** Who invented the **buoy** history **has** failed **to record, but it** was probably suggested **by** the piece **of** wood **or of cork** that **marked out** the position of the fisherman's **net.** For more **than a** century buoys **con-structed with** staves **of wood, and banded** with hoops of iron, **have been in use among maritime nations; but these are now in course of swift** supersession by buoys of iron or steel. **In 1845 the first iron** buoy **was sub-**mitted to the **Trinity** House **by the late Mr.** George **W.** Lenox, and since **that date important** improvements **have** been effected **in the form and** construction of buoys generally. Messrs. Brown and Lenox invented a **buoy,** ingeniously **contrived to** render **its** bell audible, even when **the buoy** itself is submerged, **the** stream of **water as it** passes through the lower **part** of the frame-work **keeping** in **motion an** undershot water-wheel, which rings **the** bell continuously.

A buoy invented **by the** late Mr. George **Herbert in 1853 is** so constructed with regard **to the** centre **of** flotation, and the **point** where the mooring-chain is **attached, that it will keep** upright **even** in a very **agitated sea.**

In 1878 the lighting of buoys with compressed oil **gas was effected by Messrs.** Pintsch. **Since** that **date the system has been developed** considerably, both **in this country and abroad; so that** these important aids to navigation are being rendered useful by night as well as by day, and therefore more valuable as accessories to lighthouses and lightships. The Pintsch gas-buoys now in use are found to burn continuously for three or six months, according to size, and to require no attention. Neither oil nor electricity has yet been applied to this purpose with success, but I think it is scarcely doubtful that electricity from storage batteries may and will be adopted satisfactorily and economically.

Automatic bell buoys, of various designs, and the Courtenay automatic whistling buoy, are found to render valuable assistance in foggy weather; but it has been remarked that none of these apparatus, unfortunately, possess that reliability and certainty of effect which should be characteristic of a coast signal. They must be used, therefore, with caution, their action being dependent on the motion of the sea surface.

We have spoken of beacons and landmarks hitherto as day-signals, but there is no reason why they should not—at all events, in many cases—be made useful also by night. The successful lighting of several by automatic apparatus in occasionally inaccessible positions— the light being furnished by electricity, compressed mineral oil gas, or petroleum spirit—is an important and a very significant fact. In 1884 an iron beacon, lighted by an incandescent lamp and the current from a secondary battery, was erected on a tidal rock near Cadiz. Contact is made and broken by a small clock,

A FLOATING BEACON OR BUOY.

which runs for twenty-eight days, and causes the light to **show a flash** of five **seconds,** followed by a total eclipse **of** twenty-five seconds. **The** clock has also a device for eclipsing the light between sunrise and sunset. This **apparatus** was the invention of Don Isas Lavaden.

In 1881, **a** beacon, automatically lighted by compressed **oil** gas, on the Pintsch system, was erected in the river Clyde; and it has had numerous successors, both in this country **and in** the United States. **In** 1881–82, **several** beacons lighted automatically by petroleum spirit on the system of Herr Lindberg and Herr Lyth **of** Stockholm were established by the Swedish **lighthouse** authorities, and they have worked satisfactorily. **In 1885,** a beacon light on this system, and another lighted by Pintsch's compressed gas, were erected by the Trinity House on the banks of the Thames, near Erith. The petroleum spirit lamp burns day and night at its maximum intensity, and shows a white light with a short occultation at periods of five **seconds.** The occultations **are** effected by **a** screen, rotated round the light **by the** ascending current of heated air from **the** lamp acting on a horizontal fan. As there is no governor, the periods of occultation are subject **to** slight **errors;** but the gas beacon, which **shows a** white flashing **light at** periods of two seconds, **is** provided with **a** clock (specially designed for this beacon), **which not only** regulates with precision the flashes **and** eclipses, but also extinguishes the light a **few minutes before** sunrise and relights **it** just before sunset, a very feeble pilot light being left burning during daylight. Arrangement is made in the clockwork for a bi-monthly adjustment to meet the lengthening or shortening of daylight. These two lighted beacons are in the charge of a boatman, who visits them at least once a week, when he cleans and adjusts the apparatus, and cleans the lantern glazing. "These systems of lighted beacons," says Sir James Douglass, "are not yet sufficiently matured for forming a decided opinion as to their relative efficiency and economy, but it may be considered certain that they will both be extensively adopted, because, in numerous cases, for the secondary illumination of ports, estuaries, and rivers, automatic beacons can be installed to meet fairly the local requirements of navigation, at a fraction of the first cost and annual maintenance of a lighthouse with its keepers and accessories."

Buoys have been constructed as much as twenty feet in length, and as little as four feet, but the dimensions now generally adopted range between six and thirteen. As they vary in size, so they vary in colour; usually the buoys in a river channel are painted red, striped with white, if the homeward bound vessel is to leave them on the right, and black when she has to pass them on the left. Others are painted with horizontal stripes of red and black, or in squares and diamonds, according to the various purposes they are intended to serve. Obstacles, such as wrecks, are indicated by green buoys.

Nearly a thousand buoys are moored about the coast of England and in the tideway of her rivers. Scotland and Ireland have about two hundred each. These bear their own particular designations, forming a very diversified and somewhat amusing vocabulary. We find amongst them an " Eagle," a " Gull," a " Swallow," a " Horse," a " Mussel," a " Firefly;" also, a " Cutler," a " Constable," a " Columbine," and a " Fairy;" a " Royal Sovereign," a " Protector;" a " Tongue," an " Elbow," a " Longnose." Formerly, quite a host of different kinds of buoys was in favour, and each kind had its distinctive name. As thus:— Run Buoys, Cone Buoys, Cone Reversed, Keel Cone, Egg Bottom, Hollow Bottom (Herbert's patent), Flat Bottom, Convex Bottom (Poulter's), Keel Buoys, and Spiral Buoys. But a Trinity House committee, in 1883, decided in favour of " cone " and " conical " buoys, as the most convenient for adoption as contrasting shapes; further, that middle grounds occurring in a channel, or which may divide two channels, should have at each end a spherical buoy; and, finally, that outlying dangers or positions requiring an extraordinary mark should be indicated by pillar or spar buoys.

The reader's attention must be directed, in the last place, to the modes of signalling adopted along our coasts in foggy and heavy weather, when lights are unavailing, or too vaguely defined to be regarded as safe guides. It is obvious that such modes must appeal to our sense of hearing; and that the principles governing them must depend on the greater or less facility with which sound is conveyed by the atmospheric medium, and the degree **of** ease with which the human ear appreciates and distinguishes it. For a long time it was believed that fog and **snow** exercised **a** deadening effect on sound; but it **has** been proved by experiment that **this is** not the case, at **least** to any considerable extent, and that, **in truth,** sound is assisted rather **than** obstructed **by a moist** atmosphere. During some experiments at the South Foreland, it was demonstrated that a certain atmospheric clearness is not a good medium for either sound or light. It was observed **that** singular changes in the transparency of the air occurred without any visible haze or mist intervening between the **eye** and the lighthouse towers. Sometimes **the** French lights at Calais and Cape Grisnez showed brilliantly when the photometer proved that the lights from the experimental towers, only a mile and a quarter **away,** had lost one-fourth to one-third of their power. **It is** also known that what the late Professor Henry called " the combined action of the upper and lower currents of air "—what Professor Tyndall has **described as** " water **in a** vaporous form, **mingled with air so as** to render it acoustically turbid and **flocculent," impedes the** transmission of sound; and that this " acoustic turbidity " often occurs on days **of** surprising optical transparency. Strong wind **is, of** course, a great obstructer; though instances are

on record of sound having been transmitted much
further *against* the wind than *with* it. Sometimes a
sound will be distinctly audible, then die away for
some distance, and then become audible again at a
much greater distance in the same direction. A remarkable illustration of the aberrations of sound occurred in the wreck of a vessel on the American coast.
On Little Gull Island (in Long Island Sound) is
established a powerful steam siren, which was sounding during a dead calm and dense fog, when a vessel
went on shore within a furlong of the signal station
without the signal being heard, while it was distinctly
audible at a distance of fifteen miles in another direction. Then again, much difficulty is often experienced
in " locating " the sound; that is, in determining from
what quarter it proceeds. Notwithstanding these dis-
advantages, sound-signals are of unquestionable value
as aids to navigation, and in foggy weather are absolutely indispensable. Fogs, it must be remembered,
are experienced everywhere along the coast, and in
particular localities are, at certain seasons, of daily
occurrence. The average duration in England and
Scotland is about four hundred hours annually, but at
Morecambe Bay there are nearly twelve hundred hours
of fog. At some points, however, there will be prolonged
mist in one year and very little in another; but, nevertheless, *some* fog there will most surely be, and in these
days of quick steamers, it is essential that fog-signals
should be everywhere, and everywhere efficient.

The earliest sound signal, if we **accept** the old legend, was the bell **on the** float which **the Abbot of Arbroath placed over the** Inchcape Rock. **Bells have frequently been** employed **as** fog-signals **at the entrances to harbours, and at rock** lighthouses, as, **for** instance, at the Eddystone. **In 1811 Mr.** Robert Stevenson provided the **Bell Rock tower** with a couple of sonorous bells. These are **struck** every half-minute by the machinery **which causes the rotation** of the illuminating apparatus. **His example has** been very generally followed. **The bells are struck by the hammer** from outside, **instead of by a clapper from the inside,** and **a more powerful sound is thus secured. The range depends, of course, on** the size of the **bell, and here a limit is** fixed by inflexible conditions. **Probably** the bells placed at the **new** Eddystone, **which weigh** two tons each, are as **heavy as can safely be used. The** Scottish lighthouse engineers **have found that the sound** is transmitted farthest when **the bells are** struck in rapid succession ; and **therefore,** at **the Dhu** Heartach Lighthouse, they **are struck very quickly for ten seconds, and are then silent for half a minute. This** arrangement **is effective. But in inland waters frequented by** shipping carrying **bells of their own, a new distinction** becomes necessary, and **at** Fort Matilda, **near** Greenock, Messrs. Stevenson have **therefore adopted the device of two bells of** different **tones, which are worked** by gas engines with great **success. Bells** have also **been attached** to buoys, **and are much used on the** English coast. Such bells weigh about **eight** hundredweight, and are struck **by** hammers hanging **outside,** the hammers being set in motion **by** the movement of the waves. **The** sounds **thus** obtained are audible only at short distances.

Lightships are **now being** equipped with powerful fog-signal apparatus, which **will** supersede the traditional gongs. Guns are also **made use** of, and during **fogs are** generally fired **every** ten or fifteen minutes, each discharge **costing two** shillings. Recently, a gasgun has been invented by **Mr. J. R.** Wigham **of** Dublin, **which, for signalling purposes,** possesses several advantages—**whenever, that is, a supply of** gas is available. It **can be loaded and fired at** a considerable distance **from the point** of explosion, which insures **the safety of the men who** discharge it; **and** it can be applied **when wanted** without a **moment's** delay— **a** desideratum **of** considerable importance when a ship is running into **danger. The** so-called gun is, in reality, **a tube** about twelve feet long and eighteen inches **bore, fixed** where **the** signal is required to be **made, and connected with the gas** supply by iron piping. It is loaded with **an** explosive **mixture of gas and** atmospheric air **by** simply turning on **a tap, and fired** by a **light** applied by percussion or otherwise **to** one end **of the** tube, the explosion taking place at **the mouth of the gun** almost instantaneously. An **invention of so much** ingenuity and efficiency ought **to come into very general** use.

A new form of rocket, the "**sound** rocket," has been

introduced since 1874, when experiments at Woolwich Arsenal demonstrated the superior value of gun-cotton as an explosive, the detonation of half a pound of gun-cotton giving a result fully equal to that of the firing of a three-pound charge of gunpowder. The rocket is thus composed :—1. The rocket itself is a case charged with the ordinary rocket composition, and is intended merely to carry up the explosive charge to the required height, and then to ignite the detonator which is to explode the tonite. 2. The detonator is an enlarged percussion cap, filled with fulminate. Its duty is to cause an explosion to take place in the heart of the tonite charge. The detonator is ignited by the burning of the rocket composition. 3. The tonite cartridge. This is the explosive which produces the report, and, with the detonator placed inside it, is to be fitted in the head of the rocket when required for immediate use. The adjustment of the three parts can be accomplished in less than a minute. The rocket is then lighted by the application of an ordinary fusee to a piece of Bickford fuse, communicating with the rocket composition. In less than two minutes the whole operation is effected ; the explosive charge mounts to a height of six hundred feet, and bursts with a loud report. The cost of the rocket is about one shilling and fourpence. It is cheaper, therefore, than the gun, and has the advantage of being adapted for use in the limited space offered by a rock lighthouse station. It has been furnished to the lighthouses at Lundy Island, Flamborough Head, the **Smalls,** the South Bishop Rock, the Tuskar Rock, and Heligoland.

Another form of explosive signal is **fired** by electricity, the charge with the detonator being suspended from the end of a portable **arm** which is projected from the lighthouse tower. This is adopted **at** the Longships Lighthouse, and on **board** the Breaksea Lightship; but great precautions **are necessary to** prevent accidents, **and we do not anticipate that it will** ever come into **very** general **use.**

Some eighty years ago, Mr. Robert Stevenson, as the result of some experiments on the effect of sound during **fog,** came **to the conclusion** that the "tremulous" **and sustained noise** produced **by a horn** or bugle is preferable **to that of** a bell or even **a** gun. To the same conclusion the American and other Lighthouse Boards **have** arrived. **Thirty-five** years ago, Mr. C. L. Daboll, **an American,** introduced a reed horn, or trumpet, with a metallic reed, sounded by compressed **air, and at** the suggestion of Messrs. Stevenson it was tried at the Cumbrae Lighthouse in 1865. This signal has been heard during fog at a distance of seven to nine **miles** against a light breeze, and at greater distances down the wind. It has proved of immense value **to ships** navigating the estuary of the Clyde, enabling them **to proceed to their** destination during thick fogs. So complete has been **its success that it has** been introduced at several **other** stations in Scotland.

There are now five fog-signal stations in the estuary of the Clyde, besides one worked by hot-air engines on Pladda Island; one at St. Abb's Head, Mull of Kintyre, and Sanda, each worked by hot-air engines; and at Ailsa Craig and Langness, worked by a gas engine. At all these lighthouses the "siren," the most powerful sound-signal known, has been adopted. It was patented, before 1872, by Messrs. Brown of Progress Works, New York, and consists of a long cast-iron trumpet, into the throat of which is fitted a fixed flat disk, the other end of the trumpet being connected with a steam or an air-pipe. The disk has twelve radial slits, and behind it is a rotating disk with twelve similar slits, the rotation being accomplished by separate mechanism. As one disk is fixed and the other revolving, it is obvious that the slits of each disk will frequently coincide; and as they coincide a blast of air, compressed to twenty pounds on the square inch, escapes into the siren and produces a piercing sound of unmistakable force and distinctness. "The sound of the St. Abb's signal can be heard at Crail—a distance of twenty-six miles—but this is only under peculiarly favourable conditions. No mariner could, however, depend on hearing this signal at such a distance, and, indeed, it has been laid down as a maxim in coast-signalling that the minimum and not the maximum range is what must be relied on. This minimum distance is between two and three miles: want of attention to this point may be followed by disastrous consequences."

Great improvements have been introduced into the mechanism of this modern "siren," which, unlike the siren of the old mythology, instead of luring the mariner to his destruction, seeks to compass his safety. The patent siren of Professor Holmes consists of two cylinders with angular slots, one cylinder being fixed, and the other left free to revolve within it; the compressed air impinging against the inclined sides of the slots causes the inner cylinder to revolve, so that the rapid passage of one row of slots over the other produces a series of vibrations which yield the desired note. Mr. Slight and Sir James Douglass have also effected improvements, which increase the intensity of the sound; and Messrs. Smetter, Lemonnier, & Co., have invented a double siren, in which two sirens, with different numbers of orifices in their respective cylinders, simultaneously produce two notes in the trumpet, thereby increasing its usefulness very considerably. At one station, for example, two blasts in quick succession are given, the first a low note, and the second a high note, the blasts being repeated at intervals of four minutes. At another station the signal is arranged to give blasts of five seconds' duration, with silent periods of forty seconds. In this way the variations are secured which help to indicate to the sailor the particular part of the coast he is approaching when the voice of the siren strikes upon his attentive ear.

CHAPTER XII.

LIFE IN THE LIGHTHOUSE.

IT is impossible to deny that the life of a lighthouse-keeper is distinguished by a somewhat sombre monotony; yet, on the other hand, to a serious mind, it must be cheered and elevated by the reflection that it is devoted to a high and holy service. An almost heroic simplicity invests it with peculiar dignity; it is set apart, in no inconsiderable degree, from the commonplace aims and concerns of the work-day world; it is pervaded, moreover, by an air of moderation, self-control, and almost of austerity, which resembles the lives formerly led in grotto and cavern by saint and hermit. The light-keeper is of the world, but not in it; and one would suppose that in the dwellers on the Eddystone or the Bell Rock the contests, the rivalries, the pleasures, the interests of work-day existence awake no emotion or anxiety.

The first article of the instructions which every light-keeper is bound to obey—and to obey as implicitly as a soldier obeys the laws of military discipline—run thus:

"You are to light the lamps every evening at sun-setting, and keep them constantly burning, bright and clear, till sun-rising."

This is the primary condition of a light-keeper's duty; for this he lives, toils, and watches, in order that the warning light, which has been the salvation of so many tall ships and their gallant crews, may burn with uninterrupted and steadfast ray through the hours of darkness. "Whatever else happens," says a graphic writer, "he is to do this. He may be isolated through the long night-watches, twenty miles from land, fifty or a hundred feet above the level of the sea, with the winds and waves howling round him, and the sea-birds dashing themselves to death against the gleaming lantern, like giant moths against a candle; or it may be a calm, voluptuous, moonlight night, the soft air laden with the perfumes of the Highland heather or the Cornish gorse, tempting him to keep his watch outside the lantern, in the open gallery, instead of in the watch-room chair within; the Channel may be full of stately ships, each guided by his light, or the horizon may be bare of all signs of life, except, remote and far beneath him, the lantern of some fishing-boat at sea. But whatever may be going on outside, there is within for him the duty, simple and easy, by virtue of his moral method and orderly training, 'to light the lamps every evening at sun-setting, and keep them constantly burning, bright and clear, till sun-rising.'"

That this great article of the lighthouse-keeper's

faith may be the more easily carried out, he is subjected, both when on probation and afterwards, to a strict discipline, and is required to gain a thorough acquaintance with all the materials he has to handle—lamps, oil, wicks, lighting apparatus, and revolving machinery. Before being admitted into the service, he is carefully examined as to his physical qualities by keen medical eyes; and as to his moral qualities, the best testimonials are necessary from persons in whose competency and honesty of judgment implicit confidence can be placed. He receives liberal wages, and, when past work, a fair pension; and a deduction from his pay is regularly applied to the discharge of a premium on his life insurance. He is enjoined to "the constant habit of cleanliness and good order in his own person, and to the invariable exercise of temperance and morality in his habits and proceedings, so that, by his example, he may enforce, as far as lies in his power, the observance of the same laudable conduct by his wife and family." The utmost vigilance is expected of him when it is his turn to attend to the lantern. "He whose watch is about to end is to trim the lamps, and leave them burning in perfect order, before he quits the lantern and calls the succeeding watch; and he who has the watch at sunrise, when he has extinguished the lamps, is to commence all necessary preparations for the exhibition of the light at the ensuing sunset." No bed, sofa, or other article on which to recline, is permitted, either in the lantern or in the apartment under the lantern known as the watch-room.

From these requirements we may infer what kind of life is led by the lighthouse-keeper, and what are its leading requisites—temperance, cleanliness, honesty, conscientiousness, zeal, watchfulness. At different stations it varies considerably in its lighter occupations. In the rock lighthouse—such as the Eddystone—the keeper's chief amusements are necessarily reading and fishing: the only capability of exercise is within the circle of the outer gallery, or on the belt of rock surrounding the lighthouse base: and the sole incidents which break up the uniformity of his daily life are the inspections of the committee, the visits of the district superintendent, or the monthly relief which takes the men back to shore. In the shore lighthouse—as at Harwich or the Forelands—there is a plot of ground to cultivate, frequent intercourse with visitors from the neighbouring watering-places, and the wider range of occupation and entertainment which necessarily can be enjoyed upon *terra firma.*

As a rule, the public take but little interest in the economy of our lighthouses; and yet there is something singularly romantic in the idea of the lone tower encircled by boiling waters, with its warning light flashing through the deep night shadows, and the heroic men who hour after hour watch with anxious care lest its radiance should be obscured or extinguished.

" And as the evening darkens, lo ! how bright,
   Through the deep purple of the twilight air,
Beams forth the sudden radiance of its light
   With strange, unearthly splendour in its glare !

"Not one alone : from each projecting cape
   And perilous reef along the ocean's verge
Starts into life a dim, gigantic shape,
   Holding its lantern o'er the restless surge.

"Like the great giant Christopher it stands
   Upon the brink of the tempestuous wave ;
Wading far out among the rocks and sands,
   The night-o'ertaken mariner to save.

" And the great ships sail outward and return,
   Bending and bowing o'er the billowy swells ;
And ever joyful, as they see it burn,
   They wave their silent welcomes and farewells." *

As a proof of the romance that formerly invested lighthouse life, we may lay before the reader one or two " true stories."

Smeaton speaks of a shoemaker who entered the Eddystone Lighthouse because he longed for a solitary life ; he found himself less a prisoner on his wave-beaten rock than in his close and confined workshop. When some of his friends expressed their astonishment at his choice—" Each to his taste," said he ; " I have always been partial to independence."

Perhaps it was the same individual who, after having served at the Eddystone upwards of fourteen years, conceived so strong an attachment to his prison that for two consecutive years he gave up his turn of relief. He would fain have continued the same course of

* Longfellow.

life for a third year, but so much pressure was brought to bear upon him that he consented to avail himself of the usual privilege. All the years he had spent in the lighthouse he had been distinguished for his quiet and orderly behaviour; on land he found himself "out of his element," and drank until he was completely intoxicated. In this condition he was carried back to the Eddystone, where, after languishing for a few days, he expired.

Some men have gone mad, or nearly so, by dint of contemplating the same scenes and the same external impressions. About a mile and a quarter from the Land's End, on a group of granite islets washed by the sea, stands the Longships Lighthouse, constructed in 1793. The particular rock on which it is built—the Carn-Bras—rises about forty-five feet above the level of low water. In winter both the rock and the building—as is the case at the Eddystone—will sometimes be covered for a few seconds by the leaping waters, which have even been known to surmount the lantern, and, on one occasion at least, to break through its crystal walls and extinguish the lamps.

One day, in 1862, two black flags floated from the summit of the tower. They were evidently intended as a signal of distress. What, then, had happened?

Of the three men who inhabited the lighthouse, the one whose turn it was to keep watch had thrust a knife into his breast. His companions attempted to stanch the blood by plugging up the wound with bits of tow.

Three days passed by before the people on shore could reach the lighthouse ; and the sea was then so rough and disembarkation so dangerous that the wounded man had to be lowered into the boat, suspended from a kind of impromptu crane. When he was conveyed ashore he received every attention which his condition demanded, but he lived only a few days. The jury, acting upon the evidence of his companions, declared that he had committed suicide under an attack of temporary insanity. Perhaps it is not astonishing that persons of a susceptible or excitable temperament should, under the influence of ever-murmuring seas and ever-blowing winds, and while living in a state of almost continual solitude and comparative monotony, feel the vertigo of the abyss ascend to their brain, so that the control of reason is loosened, and the mind yields to the first impulse which passes over it.

THE END.

www.ingramcontent.com/pod-product-compliance
Lightning Source LLC
Chambersburg PA
CBHW021940220326
41599CB00011BA/928